T0338646

The Grand Designers

The airplane has experienced phenomenal advancement in the twentieth century, changing at an exponential rate from the Wright brothers to the present day. In this ground-breaking work based on new research, Dr. John D. Anderson, Jr., a curator at the National Air and Space Museum, analyzes the historical development of the conceptual design process of the airplane. He aims to answer the question of whether airplane advancement has been driven by a parallel advancement in the intellectual methodology of conceptual airplane design. In doing so, Anderson identifies and examines six case histories of "grand designers" in this field, and challenges some of the preconceived notions of how the intellectual methodology of conceptual airplane design advanced. Filled with over one hundred illustrations which bring his words to life, Anderson unfolds the lives and thoughts of these grand designers.

DR. JOHN D. ANDERSON, JR. is currently Curator for Aerodynamics at the National Air and Space Museum, Smithsonian Institution, and Professor Emeritus at the University of Maryland. Dr. Anderson has published eleven books including *A History of Aerodynamics* (Cambridge University Press, 1999). He is a member of the National Academy of Engineering, an honorary fellow of the American Institute of Aeronautics and Astronautics, and a fellow of the Royal Aeronautical Society.

Cambridge Centennial of Flight

General Editors

John D. Anderson, Jr., *Curator of Aerodynamics, National Air and Space Museum, and Professor Emeritus, Aerospace Engineering, University of Maryland*
Von Hardesty, *Smithsonian Institution*

The series presents new titles dealing with the drama and historical impact of human flight. The Air Age began on December 17, 1903, with the epic powered and controlled flight by the Wright brothers at Kitty Hawk. The airplane rapidly developed into an efficient means of global travel and a lethal weapon of war. Modern rocketry has allowed heirs of the Wrights to orbit the Earth and to land on the Moon, inaugurating a new era of exploration of the solar system by humans and robotic machines. The Centennial of Flight series offers pioneering studies with fresh interpretive insights and broad appeal on key themes, events, and personalities that shaped the evolution of aerospace technology.

Also published in this series

Jeremy R. Kinney, *Reinventing the Propeller: Aeronautical Specialty and the Triumph of the Modern Airplane*
Von Hardesty, *Camera Aloft: Edward Steichen in the Great War*
Asif A. Siddiqi, *The Red Rockets' Glare: Spaceflight and the Russian Imagination, 1857–1957*
Michael B. Petersen, *Missiles for the Fatherland: Peenemünde, National Socialism, and the V-2 Missile*
Scott W. Palmer, *Dictatorship of the Air: Aviation Culture and the Fate of Modern Russia*

The Grand Designers

The Evolution of the Airplane in the 20th Century

JOHN D. ANDERSON, JR.
Smithsonian Institution

and

University of Maryland

Shaftesbury Road, Cambridge CB2 8EA, United Kingdom

One Liberty Plaza, 20th Floor, New York, NY 10006, USA

477 Williamstown Road, Port Melbourne, VIC 3207, Australia

314–321, 3rd Floor, Plot 3, Splendor Forum, Jasola District Centre, New Delhi – 110025, India

103 Penang Road, #05–06/07, Visioncrest Commercial, Singapore 238467

Cambridge University Press is part of Cambridge University Press & Assessment, a department of the University of Cambridge.

We share the University's mission to contribute to society through the pursuit of education, learning and research at the highest international levels of excellence.

www.cambridge.org
Information on this title: www.cambridge.org/9780521817875

DOI: 10.1017/9780511977565

First published 2018

A catalogue record for this publication is available from the British Library

Library of Congress Cataloging-in-Publication data
NAMES: Anderson, John D., Jr. (John David), author.
TITLE: The grand designers : the evolution of the airplane in the 20th century / John D. Anderson, Jr., Smithsonian Institution and University of Maryland.
DESCRIPTION: New York : Cambridge University Press, 2018. | Series: Cambridge centennial of flight | Includes bibliographical references.
IDENTIFIERS: LCCN 2017039156 | ISBN 9780521817875 (hardback)
SUBJECTS: LCSH: Airplanes–Design and construction–History–20th century.
CLASSIFICATION: LCC TL671.2 .A58 2018 | DDC 629.134/10922–dc23
LC record available at https://lccn.loc.gov/2017039156

ISBN 978-0-521-81787-5 Hardback

Contents

Preface

The evolution of the airplane is one of the most important technical developments of the twentieth century, and the evolution of the intellectual methodology for airplane design is part of this development. This book is focused on that intellectual methodology, how it came about, and some of the key people who advanced the methodology. It is a story of rapidly advancing technology, used ingeniously by a few people, all of whom were different. They had various backgrounds and different personalities, but fashioned out of whole cloth some of the most spectacular airplanes in history. How did they do it? This book is aimed directly at readers, both nontechnical and technical, who want to know the answer. If you are simply interested in airplanes and have a nontechnical background, this book is for you. If you are an engineer or scientist, this book is also for you. This story is one of the most fascinating in the history of technology, and it is built around two basic thoughts, as follows.

Thought one: The design of a new airplane starts by someone or some group taking out pages of blank paper (or a blank computer screen), and beginning an intellectual process called *conceptual* airplane design. After a very short period of time (weeks, or at most months), a crude configuration layout emerges showing the overall shape and size of the new airplane. After this baby, so to speak, is born, it enters a much more sophisticated and complex phase called preliminary design, where only relatively minor changes are made to the configuration layout; a lot of attention is paid to structural and control system analysis. At the end of this preliminary design phase, a major decision is made whether or not to commit to the manufacture of the airplane. If the decision is "go," the detail design phase starts – the "nuts and bolts" phase that readies the airplane for fabrication. But it is the "baby stage," the conceptual design phase on those first few pages, which dictates the very genes of the airplane.

Thought two: Anyone looking at the progress made in the advancement of the airplane sees an exponential change over the past 100 years. Just compare the image of the 1903 Wright Flyer, flying at 30 mph at about a 10-foot altitude, with the image of the spectacular Lockheed SR-71 Blackbird flying at Mach 3 plus, at 90,000 feet, or even the image of the Airbus 380 carrying over 600 passengers at near Mach One at 40,000 feet. Clearly, the airplane has experienced a phenomenal advancement in the twentieth century.

Question: Why? In particular, since the first step in the intellectual process in airplane design is conceptual design, has this exponential advancement in the airplane been due to an exponential improvement in the intellectual methodology of conceptual airplane design? This book is devoted completely to an investigation of the answer to that question. It looks at the historical development of the design process and uses six case histories of specific airplane designers, starting with the Wright brothers. I have identified these six designers as "grand designers" because they were all exceptional and different, and they make this history come alive. They are not by any means the only grand designers, but they suffice to help answer the question.

The answer itself turns out to be quite surprising, at least to me. As I progressed through the case histories, my preconceived perception of how conceptual design advanced was totally turned on end. Continue to read on; the lives and design thinking of these grand designers will unfold for you.

I wish to give credit to my colleagues at the National Air and Space Museum for many stimulating conversations on the subject of the history of aeronautical engineering, and to my wife Sarah-Allen Anderson for living with my research and thoughts on this book. Credit also goes to Dr. Von Hardesty, Senior Curator and Chair of the Aeronautics Department at the National Air and Space Museum (now retired) for suggesting the general idea of this study. Also, thanks go to Brian Riddle, Chief Librarian of the Royal Aeronautical Society in London. I spent many weeks in this library researching the topics in this book, and Brian was indispensable to me in finding material very relevant to my study. Finally, I give thanks to my longtime friend and scientific typist Susan Cunningham for typing the manuscript.

1

Introduction

Before the Beginning

Failures, it is said, are more instructive than successes; and thus far in flying machines there have been nothing but failures.

Octave Chanute, *Progress in Flying Machines*

The scene: George Cayley's Brompton Hall estate in Yorkshire, England. *The time*: one day in 1804 (the precise date is unknown). *The character*: a relatively young man of 31, carrying a wooden rod about a meter in length with a kite-like paper wing attached near the front end, and a cruciform paper tail at the back end (Figure 1.1). Sixth in a long line of baronets at Brompton Hall and self-educated particularly in scientific and technical matters, Sir George Cayley had been interested in mechanical flight for more than a decade. The object in his hands was of his own making, a small glider that embodied Cayley's seminal thinking about flying machines. Four years earlier, in what can be described only as a stroke of genius, Cayley conceived the modern configuration airplane – a flying machine consisting of a fixed wing attached to a fuselage, with a tail in the back. The sole purpose of the fixed wing was to produce lift, sustaining the aircraft as it flew through the air. In turn, Cayley's concept called for a completely separate mechanism to produce the forward thrust required to overcome the aerodynamic drag holding the aircraft back during its motion through the air. This idea completely overturned the prevailing wisdom that both lift and thrust should be generated simultaneously from a set of flapping wings trying to emulate bird flight. *The action*: Standing at the top of a hill, Cayley hand-launched his glider into the air and watched with pleasure as it glided silently and gracefully down the slope of the hill. The wing provided the lift, and the force of gravity

FIGURE 1.1. George Cayley's sketch of his 1804 glider.

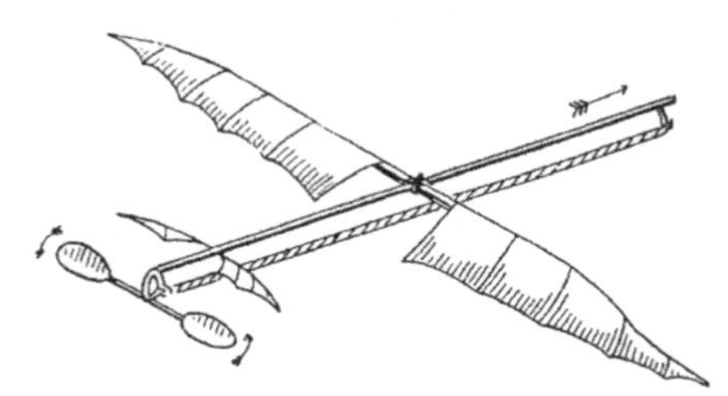

FIGURE 1.2. Alphonse Penaud's rubber-powered Planophore, 1871. Sketched by Octave Chanute in 1894.

pulling the glider down the hill was in this case, the separate mode of propulsion. In that first instant in 1804 when Cayley hurled his glider into the air, the modern configuration airplane took flight for the first time. It set the stage for the Wright brothers' successful invention of the first practical airplane a century later, and for the evolution and revolutions in airplane design to follow in the twentieth century.

The scene: the Tuileries Gardens in the center of Paris. *The time:* Friday, August 18, 1871. *The characters*: Alphonse Penaud, a young Frenchman, and a small group of colleagues united by interest in flying machines. Penaud held in his hands a small wooden and cloth model airplane, about 20 inches in length with a fixed wing near the front, a horizontal and vertical tail near the back, and a propeller at the very end (Figure 1.2). The propeller was driven by a rubber band that stretched the length of the rod-like fuselage. The Tuileries were not a pleasant sight at that time, bordered by scenes of devastation from France's defeat in the Franco-Prussian War just a year prior, and ravaged by subsequent civil warfare. But on that Friday, Penaud and his friends were oblivious to such matters, concentrating only on the model aircraft in his hands. Penaud designed the wings with a marked amount of dihedral, a geometric feature where each wing is bent upward through an angle at the fuselage. When viewed from the front of the plane, the wings formed a V-shape. Penaud intended the dihedral to provide inherent lateral stability – stability that would resist any undesirable rolling motion of the aircraft. Furthermore, he set the horizontal tail at a small negative (nose-down) angle to the

FIGURE 1.3. Hiram Maxim and his large flying machine, 1894. (National Air and Space Museum, SI-97-15315.)

fuselage to obtain proper longitudinal balance, so that the aircraft would fly in a more or less level path, rather than nosing up or down. *The action*: Ready for the test flight, Penaud wound the rubber band by turning the propeller through 240 revolutions. Holding the model at head height, he let go. The small aircraft flew beautifully, covering about 40 meters over the ground in 11 seconds, becoming the first flying machine to successfully exhibit the essential features of inherent stability. With this, Alphonse Penaud added the implementation of inherent stability to Cayley's modern configuration airplane; Penaud's design features have carried through to the present day.

The scene: Baldwyns Park, Kent, England. *The time*: July 31, 1894. *The characters*: Hiram Maxim, a self-made inventor, made wealthy by his invention of the first machine gun in history and a group of technicians and helpers. In 1888, a group of businessmen challenged Maxim to design a flying machine. Thinking big, Maxim responded, designing and building the gigantic aircraft shown in Figure 1.3. It weighed 8,000 pounds, and had a very large wingspan of 104 feet (tip to tip); the total area of the lifting surfaces was 4,000 square feet. The machine was powered by two enormous propellers, each with a diameter of 17 feet, 10 inches, which were connected to two very efficient, lightweight steam engines of Maxim's design, producing a total of 362 horsepower. Maxim mounted his machine on dual railway tracks for takeoff. Interested only in proving that a large flying machine could be designed that would have enough power and lift to get off the ground and sustain itself in flight, Maxim used a guard rail that limited the machine from climbing any higher than 2 feet off the ground. *The action:* Maxim and two other crew members climbed aboard. He applied full power and the machine accelerated down the rails, ran 600 feet, and then almost effortlessly lifted off the track. It smashed through its restraining rail and floated free, "giving those on board the sensation of being in a boat" as stated by Maxim. He immediately cut off the steam, and the huge machine settled gently back to the ground. By then it had covered over 1,000 feet in the air. Maxim never flew again, but he had demonstrated that a large heavier-than-air flying machine could generate enough lift to leave the ground under its own power. Suddenly, the day of the flying machine equipped with a prime mover with enough power (enough "steam") to get off the ground had dawned.

The scene: a houseboat moored off Chopawamsi Island near the western bank of the Potomac River at Quantico, Virginia, more than 3,000 miles to the west of Baldwyns Park, across the Atlantic Ocean. *The time*: about 3:05 pm on May 6, 1896, two years after Maxim's powered liftoff. *The characters*: Samuel Pierpont Langley, third secretary of the Smithsonian Institution in Washington, D.C., accompanied by Alexander Graham Bell, his close friend. *The action*: Slung underneath a catapult mounted on the roof of the houseboat was a flying machine in the form of two equal sized rectangular wings, placed one behind the other (a tandem wing arrangement). Made from spruce and covered with China silk, each wing measured 13.1 feet from wing tip to wing tip. A horizontal and vertical tail was located at the rear of the machine. Between the two wings were dual propellers powered by a single lightweight one horsepower steam engine. The machine was too small to carry a pilot but too large to be considered a model. This was a serious flying machine designed by Langley and based on seven years of painstaking aerodynamic research carried out on a large whirling arm device. It was also based on observations of the flight of various rubber band-powered model aircraft that Langley tossed out of windows of the Smithsonian's castle-like building in

FIGURE 1.4. Langley's 1896 steam-powered aerodrome seen in a photo taken by Alexander Graham Bell.
(National Air and Space Museum, NAM-A-12582.)

a large open area in downtown Washington. Langley called his flying machine an "aerodrome" (terminology based on his misinterpretation of a Greek term for such an object). On that day, the aerodrome was launched into a gentle breeze from a height of 20 feet above the river's surface; initially it slowly descended about 3 or 4 feet, then began to climb steadily. Bell excitedly took a photograph of the aerodrome moments after launch (Figure 1.4). Bell was present because of his intense interest in powered flight, in addition to his close friendship with Langley. The aerodrome ultimately reached a height of 100 feet before it ran out of steam. By the time it had settled gently in the water, it had been in the air for one and a half minutes and had covered a distance of 3,300 feet.

The flight was momentous. It was the first truly successful sustained flight of a heavier-than-air powered machine in history. The earlier flight of Penaud's rubber band-powered model and Maxim's gargantuan machine pale in comparison. Langley had demonstrated to the world the technical feasibility of such powered flight beyond a shadow of a

FIGURE 1.5. Otto Lillienthal flying one of his gliders in 1893. (National Air and Space Museum, SI-87-17029.)

doubt. This demonstration was not lost on Orville and Wilbur Wright, who subsequently took the technical feasibility essentially for granted.

Powering a machine into the air is one thing: flying such a machine is quite another. At the same time that Langley launched his successful steam-powered aerodrome, Otto Lilienthal in Germany enjoyed successful flights in gliders of his own design. Lilienthal developed a passion for flying in 1861 at the age of 13. A mechanical engineer with a degree from the Berlin Trade Academy (now the respected Technical University of Berlin), Lilienthal carried out seminal aerodynamic research over two decades, culminating in his design of the first successful hang gliders in history. Lilienthal's philosophy was to get into the air with gliders and learn how to fly a heavier-than-air machine before adding an engine. A photograph of Lilienthal flying one of his gliders is shown in Figure 1.5. Lilienthal's "airman's" philosophy differed from the brute-force approach of many previous flying machine inventors. Executing over 2,000 successful glider flights before his untimely death in a crash of one of his machines on August 9, 1896, Lilienthal pioneered the art of actually flying a heavier-than-air machine. With this, and with Langley's definitive demonstration of the technical feasibility of heavier-than-air powered flight, the invention of the successful airplane was just around the corner.

Missing, however, was a viable methodology for the technical design of the airplane. Would-be inventors in the nineteenth century were mainly flailing in the dark. No organized bulk of aeronautical data existed (except perhaps for the detailed aerodynamic experiments by Langley and Lilienthal). Granted, a great deal of empirical results, some useful and some not, were collected over the nineteenth century, and the basic overall configuration of the airplane (fixed wing, fuselage, and tail) became somewhat standard. Much of this was later inherited and reworked by the Wright brothers (Chapter 2). But there was no cohesive intellectual process for the design of an airplane – no agreed upon, technically sound method to even start the process. What do you do when you first sit down and begin the process of airplane design? Today, this beginning process is called conceptual airplane design. The history of the intellectual development of the process is indeed the subject of this book. People, Lilienthal and Langley included, simply used their own intuitive approaches and set forth.

The situation changed with the Wright brothers. One of the main reasons that the Wrights were the first to invent the successful airplane was that they developed a logical and technically sound methodology for airplane design.[1] We will examine the evolution of their methodology in the next chapter. They were the first of the grand designers.

Conceptual airplane design today can be defined as the intellectual process of creating on paper (or on a computer screen) a flying machine to (1) meet certain specifications and requirements established by potential users (or as perceived by the designers and manufacturers) and/or (2) pioneer innovative new ideas and technology.[2] An example of the former is the design of most commercial transports, starting at least with the Douglas DC-1 in 1932 that was designed to meet or exceed various specifications stipulated by an airline company. (The airline was TWA, named Transcontinental and Western Air at that time.) An example of the latter is the design of the experimental rocket-powered Bell X-1, the first airplane to exceed the speed of sound (October 14, 1947). Today, the conceptual design process is a sharply honed intellectual activity, a rather special one that is still tempered by good intuition developed via experience, by attention paid to past successful airplane designs, and by (generally proprietary) design procedures and databases that are part of every airplane manufacturer. Moreover, airplane designers today are much more aware of new scientific and engineering techniques and advances than in the past, and are much more inclined to incorporate new science in their design as long as that new science is reasonably proven. Throughout much of the twentieth century, however, this was not always the case. Engineering research in aeronautics has grown exponentially since the end of World War II; prior to that many advances in airplane design were

[1] Jakab, Peter L., *Visions of a Flying Machine*, Smithsonian Institution Press, Washington, DC, 1990.

[2] Anderson, John D., Jr., *Aircraft Performance and Design*, McGraw-Hill, Boston, 1999.

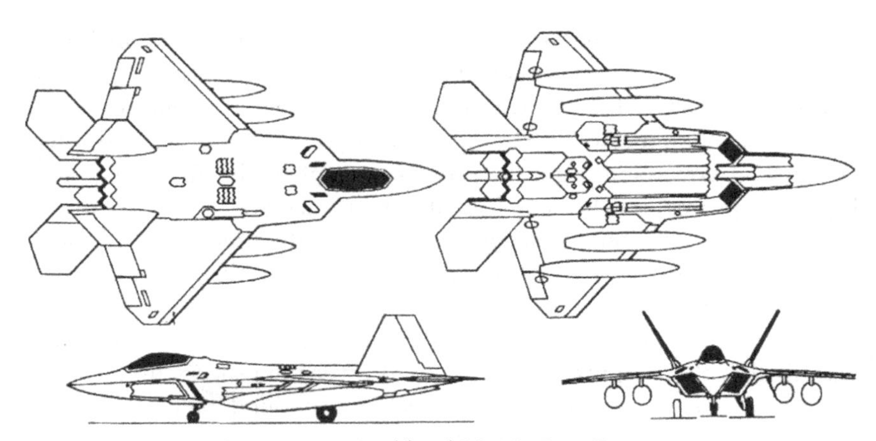

FIGURE 1.6. Lockheed Martin F-22 Raptor.

empirical and intuitive, and only later did research explain why certain things worked. Indeed, the late aeronautical historian Richard K. Smith wrote in 1989 that "the airplane did more for science than science ever did for the airplane."[3]

The overall design of an airplane today is a result of 100 years of powerful mental effort, emotional inspiration, absolute dedication, and sometimes pure luck. In particular, a mental process for the first step in this design process – conceptual design – did indeed evolve and develop over this time. The major focus of this book is the study of that intellectual evolution.

Since the beginning, people have designed airplanes. They still do, in spite of the automation and artificial intelligence brought about by modern computers. Therefore, our study of the intellectual evolution of conceptual airplane design revolves around the case histories of six successful airplane designers; we shall label them as representatives of the "grand designers." Although these grand designers are different people with different emotional and mental outlooks, we look for a common intellectual thread that ran through these people that made them successful designers. We will trace their contributions to the intellectual processes and methodology that now represent modern conceptual airplane design.

[3] Smith, Richard, K., "Better: The Quest for Excellence," in *Milestones of Aviation*, John T. Greenwood (Ed), Hugh Lauter Levin Associates, New York, 1989, p. 224.

Through World War I, when a newly designed airplane took off on its initial flight, it had about as much chance of crashing on takeoff, or failing in some other equally dramatic fashion, as it did for a successful first flight. A testimonial to the progress in airplane design is that today, crashes do not mar the first flights of new aircraft. Consider the Lockheed-Martin F-22, a modern fighter incorporating advanced technology, including stealth (Figure 1.6). The F-22 flew for the first time on December 15, 2006, staying in the air for 58 minutes, climbing to altitudes of 20,000 feet at speeds up to 300 mph, and achieving angles of attack during maneuvers up to 14 degrees. The designers did not blink an eye. The methodology of airplane design had advanced to the point where successful first flights were almost taken for granted. Airplane design has come a long way since the turn of the twentieth century. Read on for the rest of the story.

2

The Beginning

The Wright Brothers and Their Design Process

> One of the most gratifying features of the trials was the fact that all our calculations were shown to have worked out with absolute exactness so far as we can see.
>
> Wilbur Wright, in a letter to Octave Chanute

The scene: windswept sand dunes of Kill Devil hills, 4 miles south of Kitty Hawk, North Carolina. *The time*: about 10:35 am on Thursday, December 17, 1903. *The characters:* Orville and Wilbur Wright, and five local witnesses. *The action*: Poised, ready to make history, is a flimsy, odd looking machine, made from spruce and cloth in the form of two wings, one placed above the other, a horizontal elevator mounted on struts in front of the wings, and a double vertical rudder behind the wings. A 12 horsepower engine is mounted on the top surface of the bottom wing, slightly right of center. To the left of this engine a man – Orville Wright – lies prone on the bottom wing, facing into the brisk and cold December wind. Behind him rotate two ungainly looking airscrews (propellers), driven by two chain-and-pulley arrangements connected to the same engine. The machine begins to move along a 60-foot launching rail on level ground. Wilbur Wright runs along the right side of the machine, supporting the wing tip so that it will not drag the sand. About 20 feet from the end of the starting rail, the machine lifts into the air; at this moment, John Daniels of the Kill Devil Life Saving Station takes a photograph which preserves for all time the most historic moment in aviation history, shown in Figure 2.1. The machine flies unevenly, rising suddenly to about 10 feet, ducking quickly toward the ground. This type of erratic flight continues for 12 seconds, when the machine darts to the sand, 120 feet from the point where it lifted

FIGURE 2.1. Wright Flyer taking its historic first flight, December 17, 1903, Kill Devil Hills, North Carolina.
(National Air and Space Museum, NAN-A-26767-B.)

from the starting rail. Thus ends a flight which, in Orville Wright's own words, was "the first in the history of the world in which a machine carrying a man had raised itself by its own power into the air in full flight, had sailed forward without reduction of speed, and had finally landed at a point as high as that from which it started."

The machine was the Wright Flyer I that is now preserved for posterity in the National Air and Space Museum of the Smithsonian Institution in Washington, DC. The flight on that cold December 17, and the three others that immediately followed during the next one and a half hours were momentous. They brought to a realization the dreams of centuries, and gave birth to a new way of life. They were the first powered flights of a heavier-than-air, piloted, controlled machine. With them, and with the further successes to come over the next five years, came the Wright brothers' clear right to be considered the premier aeronautical engineers of history. They were the first of the "grand designers." But how did they do it? How did they pull it off when all previous would-be inventors of flying machines failed? The central theme of this chapter is to answer these questions.

PERSONAL PROFILES

Throughout this book we look for common intellectual threads that bind the grand designers – something that made them all great airplane designers although each was a different person with a different personality, from widely different backgrounds, and with diverse personal motivation. We start with Wilbur and Orville Wright. Much has been written about the lives of the Wright brothers. The story of the two men who designed and built the first practical flying machine has filled many books, among them the definitive biography of the Wright brothers by Tom Crouch, *The Bishop's Boys*.[1,2] We will delve into the personal background of the Wrights only to the extent necessary to understand the thought processes that gave birth to their design methodology.

Wilbur and Orville Wright were products of the post-civil war era. Wilbur was born near Millville, Indiana, on April 16, 1867, and Orville in Dayton, Ohio, on August 19, 1871. They grew up in the Midwest, and reflected the conservative, hardworking, independent, and can-do attitude that was prevalent there during the late nineteenth century. Seven children were born in the family; two older brothers, Reuchlin and Lorin; a set of twins who died in infancy; a younger sister, Katherine; and Wilbur and Orville. Their father, Milton Wright, was an active minister, church administrator, and eventually bishop in the United Brethren Church. Their mother, Susan Koerner Wright, was shy and scholarly, and before marrying Milton Wright she attended Hartsville College, studying literature and coming within three months of graduating. Having a college education was rare for a woman in the nineteenth century, but Susan was exceptional in many respects. She was gifted with considerable mechanical ability. She had grown up on her father's farm in Hillsboro, Virginia, and had played and worked in her father's carriage shop on the farm, where she became adept in the use of mechanical devices. What mechanical knowledge Wilbur and Orville acquired during childhood was learned from their mother. "She designed and built simple household appliances for herself and made toys, including a much-treasured sled, for her children," writes Tom Crouch. "When the boys wanted mechanical advice or assistance, they came to their mother. Milton was one of

[1] Wright, Orville, "How We Made the First Flight," *Flying*, December 1913, pp. 10–12, 35–36.

[2] Crouch, Tom, *The Bishop's Boys*, Norton, New York, 1989.

those men who had difficulty driving a nail straight."[3] Her enjoyment in working with things mechanical was absorbed by Wilbur and Orville.

The Wright brothers never officially graduated from high school. Wilbur attended high school in Richmond, Indiana, where the family was living at the time. He was an accomplished scholar, taking courses in Greek, Latin, geometry, natural philosophy, geology, and composition, earning grades well above 90 percent in them all. In addition, he was an athlete, excelling in gymnastics. He was doing so well that Milton and Susan were considering sending him to Yale. Unfortunately, just before Wilbur was to graduate in June 1884, Milton abruptly moved the family to Dayton because of compelling church and political reasons. Wilbur was not able to complete the courses required for graduation. Consequently, he never received a high school diploma and never attended Yale. He continued his own education, however, ultimately becoming more well-read than most college graduates. Orville attended Central High School in Dayton. Unlike Wilbur, Orville was not an outstanding student, but he managed to earn grades in the 70–90 percent range. Orville took several advanced courses in his junior year that were not part of the required curriculum, and as a result he was going to be several credits shy of the number necessary for graduation. Because he would not be able to graduate with his classmates, Orville decided not to return. His mother died just before he would have returned for his senior year, reinforcing his decision. In short, the inventors of the first practical flying machine never received high school diplomas, although they proved to be better educated than most.

Orville immediately satisfied his mechanical and business aspirations by going into the printing business. He built several successful printing presses of his own design, constructed from a collection of spare parts. In 1889 he began publishing a reasonably successful local, weekly newspaper, *The West Side News*. Within months he managed to draw Wilbur into the business. Wilbur expanded the newspaper into a daily entitled *The Evening Item*, promising the readers "the clearest and most accurate possible understanding of what is happening in the world from day-to-day." At that time, however, Dayton had twelve newspapers, some with new high-speed presses that simply outclassed the Wrights. On a limited budget, and with the thought of going into debt simply not part of their character, they ceased publication in August 1890.

[3] Ibid, p. 33.

However, they continued with job printing until 1892, when they both simply lost interest, and focused on something new – bicycles. Their joint experience in the printing business, began a lifetime intellectual partnership that would figure strongly in their invention of the first successful airplane.

By 1892, both Wilbur and Orville were bicycle enthusiasts, competing in local races. In December, they opened a bicycle shop on West Third Street in Dayton, about two blocks from their home. At first they sold and repaired bicycles, but by 1895 they were handcrafting and selling their own line. It was a profitable business. The bicycle had become a major mode of transportation, and at the height of its popularity in the 1890s, there were more than 300 companies in the United States manufacturing over a million bicycles each year. The Wrights' talent and experience in designing and hand building bicycles helped prepare them for their later flying machine work. Peter Jakab, a leading historian of early flight and an expert on the Wrights' inventive process wrote:

> While the major bicycle manufacturers were employing mass-production techniques adopted from the firearm and sewing-machine industries, the Wrights remained small scale and continued to produce handmade originals. At a time when manufacturing was becoming increasingly mechanized and rapidly rushing towards the twentieth century, the Wrights stayed firmly within the classic artisan tradition of handcrafted, carefully finished individual pieces. This kind of attention to detail and craftsmanship would be a hallmark of their flying machines. Every component of their aircraft was designed and built with great care and served a specific and essential function. It is a bit ironic that an invention that has been so influential in the twentieth century was the product of men whose approach was so firmly anchored in the nineteenth century.[4]

By 1896 the brothers Wright were securely entrenched in the bicycle business in modest surroundings at 1127 West Third Street in Dayton and were living comfortably just three blocks away in their house at 7 Hawthorne Street. (Both the bicycle shop and the house are now located in Greenfield Village, Dearborn, Michigan, having been purchased and moved there by Henry Ford in 1936–37 as part of the museum Ford created to honor American ingenuity and industrial achievement.) Wilbur and Orville Wright (Figure 2.2) were intelligent and widely read, possessed a Puritanical moral attitude, considerable mechanical aptitude, a dedication to quality craftsmanship, and a tireless work ethic; fertile

[4] Jakab, Peter L., *Visions of a Flying Machine*, Smithsonian Institution Press, Washington, DC, 1990, p. 9.

FIGURE 2.2. Wilbur and Orville Wright on the back steps of their home in Dayton, Ohio.
(National Air and Space Museum, NAM-A-43268.)

ground indeed for nurturing the seeds of the idea of flying and meeting the engineering challenge of powered flight.

DESIGN METHODOLOGY – THE BEGINNINGS

Casually interested in the idea of fight since childhood, the Wrights read in popular literature about Otto Lilienthal's glider flights. Lilienthal's death in August 1896 had a particular effect on Wilbur, motivating him to think more seriously about heavier-than-air flight. Somewhat frustrated by the lack of substantive aeronautical literature easily available to him, Wilbur wrote to the Smithsonian Institution on May 30, 1899 requesting literature on "the problem of mechanical and human flight." He went on to state:

> I am an enthusiast, but not a crank in the sense that I have some pet theories as to the proper construction of a flying machine. I wish to avail myself of all that is already known and then if possible add my mite to help on the future worker who will attain final success.

Intuitively, Wilbur was taking the first step in the common sense evolution of his methodology for airplane design, namely, to learn what existed and what went before. The Smithsonian quickly responded by sending Wilbur several pamphlets and papers, including some written by Lilienthal and others by Langley, and most importantly for Wilbur, recommendations for the important literature on the subject. During that year, the Wrights obtained and devoured much of the recommended literature, and quickly became knowledgeable about the existing state of the art flying machines.

Simultaneously, the Wrights began to appreciate the four fundamental disciplines that are embodied in the design of an airplane – aerodynamics, propulsion, flight dynamics and control, and structures. To understand how these disciplines interact, visualize an airplane in flight. There is a rush of air over the airplane, exerting an aerodynamic force on the aircraft. The understanding of this airflow and the resultant force it exerts on the airplane is the essence of the discipline of *aerodynamics*. Part of this force is *lift* which supports the airplane in the air, countering its weight. Part of this force is *drag*, which resists the forward motion of the airplane and which must be overcome by *thrust* from a propulsive device that keeps the airplane going; the production of thrust involves the discipline of *propulsion*. The airplane moves under the action of the forces acting on it, and its resulting path and motion through the air constitutes the discipline of *flight dynamics*, which includes the stability and control of the airplane. Finally, as a result of the forces and accelerations felt by the airplane, its material structure undergoes substantial internal stress and strain. The airplane structure must be strong enough to resist these stresses and strains, and to keep from falling apart in midair. These matters constitute the discipline of *aircraft structures*. In 1899, the Wrights quickly began to appreciate the separate disciplines of aerodynamics, propulsion, flight dynamics, and structures. Moreover, the Wrights began to think of a flying machine as a system where all these disciplines must act synergistically and must perform successfully in order to achieve successful flight. In analogy with a modern stereophonic sound system, the quality of the sound is only as good as weakest component in the system. For example, an airplane may have great aerodynamics, propulsion, and flight control, but if its structure is poor, the whole system may fail catastrophically. The Wrights understood this right from the beginning. It became a cardinal principle of their design methodology: a flying machine is a *system* in which all the components must perform successfully.

The Wright brothers never wrote a book on airplane design, nor did they ever publish an article that specifically outlined their methodology in terms of a clear sequence of intellectual steps. The Wrights, however, did develop their own design methodology; it evolved systematically and naturally between 1899 and 1902 without them being fully aware that it was happening. Rather, they started out to experiment with flying machines, to "find much in advance of the methods tried by previous experimenters" as Wilbur wrote his father from Kitty Hawk on September 23, 1900.[5] Only later did they realize they were getting close to developing a successful flying machine and focused on designing one that would work. Even though they never wrote down their design methodology as such, it can be extracted from a careful reading of their voluminous correspondence and notes between1900 and 1903.[6] When examined from an engineering viewpoint, and by "connecting the dots," their writings reveal a definite evolution of a design methodology.

EVOLUTION

After digesting the available literature on flying machines, the Wrights took the next step – learning by doing. Their early thinking focused on the problem they felt was the most important to solve first – equilibrium and control of a flying machine. Sometime before July of 1899, Wilbur began to appreciate the value of lateral control, i.e., control associated with roll about the longitudinal axis through the fuselage. He also deduced that the proper way to turn an airplane is to roll the wings, tilting the lift force in the direction of the roll, thus causing the airplane to turn in the direction of the tilted lift force. Moreover, he discovered a mechanical method for achieving lateral control, namely warping the two opposite wing tips, increasing the angle of attack of one wing tip and decreasing that of the opposite wing tip. The higher angle of attack of one wing resulted in more lift on that wing, whereas the lower angle of attack of the other wing resulted in less lift, the imbalance of the two lift forces causing the airplane to roll. The concept of lateral control and a mechanical means of achieving it, is the most important contribution made by the Wrights to airplane

[5] McFarland, Marvin W., ed., *The Papers of Wilbur and Orville Wright*, Vol. I, McGraw-Hill, 1953, p. 26.

[6] Ibid. Since many references were made to this volume throughout the rest of the chapter, for practicality we dispense with detailed page references, but rather refer in the main text to the specific dates of the Wrights' correspondence where appropriate.

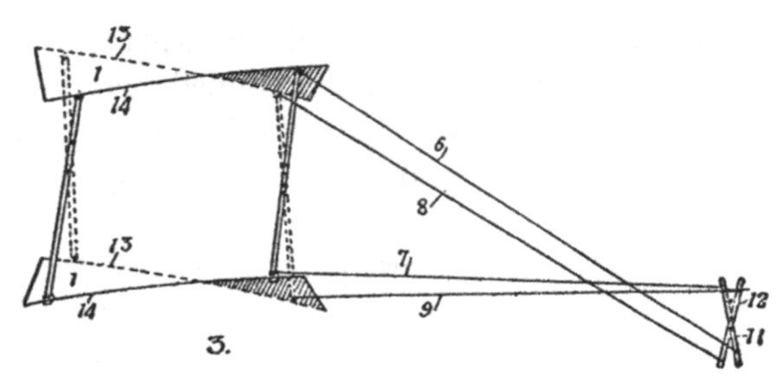

FIGURE 2.3. Wright brothers' drawing for their 1899 kite. (National Air and Space Museum, SI-2001-9902.)

design. Lateral control was not appreciated and had been overlooked in all previous designs of flying machines. In contrast, the Wrights seized on lateral control as the most important problem to face first. Their mechanical solution using wing warping was eventually replaced by the use of ailerons, but for the time being they found something that worked.

But how did they know it was going to work? Were their ideas on wing warping and lateral control viable? Before they could design any serious flying machine, they had to answer this question; they had to prove their concept. Wilbur did this by building a kite (Figure 2.3) with two wing surfaces placed one above the other – a biplane configuration. He flew the kite by manipulating strings attached to two handheld sticks.

The other ends of the strings were attached to the kite, near the tips of the wing surfaces. With the kite in flight, the wings could be warped by proper movement of the handheld sticks. As Orville wrote much later:[7] "According to Wilbur's account of the tests, the model worked very successfully. It responded promptly to the warping of the surfaces, always lifting the wing that had the larger angle." Orville was away from home at the time, and was not present for the kite flights. The precise date for the kite test is not known, but scholars place it sometime near the end of July, 1899. With these kite tests, Wilbur proved his concept of wing warping and lateral control. Moreover, he did it conclusively with a small, inexpensive model that posed no danger to human life and limb. This is in contrast to several nineteenth century inventors who designed large flying machines without proof of any concepts; for them, the proof was to be in the pudding, in the flight of the full-scale aircraft. These aircraft never

[7] Orville Wright, biographical deposition, Jan. 13, 1920, Dayton, Ohio (see McFarland, p. 11.)

successfully flew. So the Wrights' 1899 proof of concept kite broke new ground in the methodology of airplane design, and with it the Wrights began a long path of systematic learning.

Although versed in the existing literature, and encouraged by their kite test, Wilbur felt the need to consult with the leading expert on flying machines, Octave Chanute. The author of the most definite book on airplane developments at that time, *Progress in Flying Machines*,[8] Chanute played the role of the master disseminator of aeronautical information. A self-made civil engineer, he gained fame as the designer of the Union Stockyards in Chicago and Kansas City, and the first bridge across the Missouri River. He became chief engineer for the Erie Railroad, and later a consulting engineer. In 1890 he settled in Chicago, and specialized in wood preservation. Interested in aeronautics since a trip to France in 1875, he became wealthy enough by the early 1880s to devote a considerable amount of time and effort to the subject. By 1894, when his book was published, Chanute had earned an international reputation as an authority on flying machines. In 1896 and 1897 he designed a series of gliders, paid for their construction, and financed a small team to fly them from sand hills along the southern shore of Lake Michigan. Chanute became a celebrity in aeronautics, and it is no surprise that Wilbur felt compelled to consult with Chanute before he and Orville progressed any further.

On May 13, 1900, Wilbur wrote a letter of introduction to Chanute, stating in the first sentence that "for some years I have been afflicted with the belief that flight is possible to man." He went on to explain his intent to test a rather large glider by erecting a lightweight tower about 150 feet high, and flying the glider tethered from the top of the tower. He wanted to know what Chanute thought about this idea. Wilbur went on to explain his concept of wing warping for lateral control, sharing with Chanute in good faith this seminal idea. "I ... am quite in sympathy with your proposal to experiment." Chanute wrote back on May 17, and he suggested some references for the Wrights to consult. Thus began a long series of correspondence between a bicycle mechanic in Dayton, Ohio, and the world's leading aeronautical authority in Chicago, a series that would amount to over 400 letters over a span of 10 years until Chanute's death in 1910.

[8] Chanute, Octave, *Progress in Flying Machines*, 1894. Reprinted by Lorenz and Herweg, Long Beach, CA, 1976.

FIGURE 2.4. Wright brothers' number 1 glider being flown as a kite at Kitty Hawk, North Carolina, 1900.
(National Air and Space Museum, SI-2002-23711.)

On September 13, 1900, Wilbur showed up in Kitty Hawk, North Carolina, with the partially prefabricated parts for a large glider. Orville joined him fifteen days later. By the first week in October the glider made its first flights, first tethered with Wilbur on board, and later as an unmanned kite. They also tried flying it from a tower as Wilbur had first suggested to Chanute, but quickly gave up on this testing method. A photograph of the Wrights' 1900 glider flying as a kite is shown in Figure 2.4.

Take a good look at this glider. Save for two notable exceptions, the glider was off-the-shelf technology. The overall choice of the biplane configuration with rectangular wings came from earlier nineteenth century machines, and later was particularly motivated by Chanute's glider design from 1896, shown in Figure 2.5. The Wrights followed Chanute's structural technique of wing bracing by vertical struts (the Wrights called them "uprights") and diagonal wires, although they used diagonal wire bracing only along the spanwise direction, eliminating it in the fore-and-aft direction to allow for wing warping.

FIGURE 2.5. Octave Chanute's biplane hang glider, 1896. (National Air and Space Museum, NAM-A-48095-C.)

They followed Lilienthal's aerodynamics, using cambered (curved) airfoils for the wing. Lilienthal had carried out extensive aerodynamic tests on a single wing shape using a large whirling arm in the back yard of his house in Berlin. The Wrights used a different airfoil shape, however, with the maximum camber near the leading edge, in contrast to Lilienthal's circular arc airfoils with maximum camber located halfway along the chord (the distance from the leading edge to the trailing edge). In reality, at that time the Wrights were not aware of Lilienthal's airfoil shape; they had only a table of Lilienthal's normal and axial force coefficient data originally published in a handbook in Germany[9] and reprinted by Chanute in an article published in in 1897.[10] (The Wrights easily obtained lift and drag coefficients from Lilienthal's normal and axial force

[9] Hermann W. L. Moedebeck, *Taschenbuck zum Praktischen Gebrauch fur Flugtechniker und Liftschiffer*, Verlag von W.H. Kuhl, Berlin, 1895.

[10] Chanute, Octave, "Sailing Flight," in *The Aeronautical Annual*, ed. by James Means, Boston, 1897.

coefficients by simple trigonometry.) Chanute did not give the precise airfoil shape; he identified Lilienthal's airfoils as "surfaces arched upward about one-twelfth of their width," i.e., with a camber of one-twelfth. The Wrights used this copy of Lilienthal's table, and therefore did not have a clue as to the precise airfoil shape tested by Lilienthal. Nevertheless, they intuitively felt that the more forward the location of the maximum camber, i.e., the closer to the leading edge, the less would be the movement of the center of pressure (the point through which the lift effectively acts) as the angle of attack changed. They made this decision without technical proof, but rather on the basis of their common sense, as later described by Wilbur in his presentation "Some Aeronautical Experiments" to the Western Society of Engineers in Chicago on September 18, 1901.[11] Knowing that the center of pressure on a cambered airfoil moves forward as the angle of attack decreases from a high angle to a moderately low angle, and then suddenly reverses and rapidly moves rearward as the angle of attack is decreased further, Wilbur theorized in his paper that the reversal was caused by the wind striking "the forward part of the surface on the upper side instead of the lower, and thus this part altogether ceases to lift, instead of being the most effective part of all." He went on to explain that "to be on the safe side, instead of using the arc of a circle, we had made the curve of our machine very abrupt at the front, so as to expose the least possible area to this downward pressure." The Wrights used the Lilienthal table to predict the lift and drag on the wings of their glider with no regard to the airfoil shape or to the wing aspect ratio. (For a rectangular wing, the aspect ratio is the ratio of the distance along the span from one wingtip to the other, divided by the chord length, which is the distance from the leading edge to the trailing edge.) Again unknown to them at the time, Lilienthal's data were measured for a wing with an aspect ratio of 6.48; the Wrights designed their 1900 glider with an aspect ratio of 3.5. The Wrights' use of the Lilienthal table was totally inappropriate for their particular glider design.[12] Unaware, the Wrights

[11] Wright, Wilbur, "Some Aeronautical Experiments," September 18, 1901, in McFarland, ed., *The Papers of Wilbur and Orville Wright*, McGraw-Hill, New York, 1953, p. 109.

[12] New research shows that the Wrights made three separate misinterpretations of Lilienthal's data. These results are described at length by Anderson in *A History of Aerodynamics*, Cambridge University Press, 1997, pp. 205–216, *The Airplane: A History of Its Technology*, American Institute of Aeronautics and Astronautics, 2002, pgs. 100–105, and *Inventing Flight*, Johns Hopkins University Press, 2004, pp. 119, 162–164.

continued to use the Lilienthal table for the design of their 1901 glider as well, incorrectly interpreting the table as a set of universal values.

Returning to the 1900 glider shown in Figure 2.4, the two unique design features definitely *not* off the shelf are the forward horizontal canard surface ahead of the wings and the inclusion of wing warping. Both features deal with flight dynamics – the stability and control of the flight vehicle. From the 1899 kite tests, the Wrights confidently designed their 1900 glider with wing warping. The flight testing at Kitty Hawk in 1900 proved conclusively the viability of wing warping as a means to obtain lateral control, and they continued to design flight vehicles with wing warping for the next decade. By using a horizontal control surface placed ahead of the wings (canard configuration) for longitudinal (pitch) control, the Wrights deviated from the conventional nineteenth century practice of placing a horizontal tail behind the wings. This was a bold stroke, apparently based again on their intuition enhanced by fundamental thinking rather than experimental test. The Wrights do not give a detailed written description of how they came to this canard design, and the record is a bit hazy. In the process of assembling the 1900 glider, Wilber wrote a letter to his father on September 23 in which he stated: "The tail of my machine is fixed, and even if my steering arrangement should fail, it should still leave me with the same control that Lilienthal had at best." (The use of the word "tail" conjures up a surface at the rear of the machine, but no photograph or description of the 1900 glider shows a tail in the rear. Was Wilbur using a generic term just to help his father's understanding that the 1900 glider had pitch control?) In his letter of November 16 to Chanute, written from Dayton after his return from Kitty Hawk, Wilbur stated: "We laid down flat on the lower surface and maintained fore-and-aft balance by means of a forward rudder." (The Wrights referred to their horizontal surface as a "rudder.") Wilbur made no mention of ever using a tail in the rear. Later, while discussing the design of the 1900 glider in his paper to the Western Society of Engineers on September 18, 1901, Wilbur clearly states: "Our . . . work was to draw up the plans for a suitable machine. After much study we finally concluded that tails were a source of trouble rather than assistance; and therefore we decided to dispense with them altogether." Taken in context with his earlier letter to Chanute, we are led to believe that the horizontal canard surface formed an integral part of their design right from the start.

Finally, the 1900 glider differed in a third respect from previous technology. Beginning with George Cayley at the turn of the nineteenth century, most designs for flying machines incorporated wing dihedral, a slight bending up of the wings relative to the center, or root portion of the wing. A wing with dihedral, when viewed from the front, has a very shallow V-shape; a wing with no dihedral is absolutely flat from tip to tip. Dihedral provides lateral stability; a wing with dihedral, when caused to roll by some disturbance, will by itself tend to roll back toward its equilibrium position. Many flying machine developers in the nineteenth century considered lateral stability, hence dihedral, to be a necessary aspect of airplane design. The Wrights, however, eschewed dihedral. During their flight tests of the 1900 glider at Kitty Hawk, Wilbur wrote in his notebook that "a dihedral angle increases the disturbance produced by side gusts." On November 16, after his return to Dayton, Wilbur wrote a lengthy letter to Chanute describing some of the testing experience and results obtained with the 1900 glider. Among these he disclosed: "At first the machine was curved laterally to obtain the effect of dihedral angles. But we found the effect very unsatisfactory in gusty winds." This experience soured the Wrights towards dihedral, and they never again used it in their designs. This design feature was retrograde, however, because virtually all other airplane designers after the Wrights incorporated dihedral in most of their airplanes.

Although the Wrights borrowed heavily from previous nineteenth century investigators[13] they did not apply this off-the-shelf technology blindly. The technology embodied in their 1900 glider was well thought out. They based the structural and aerodynamic design on the best available data as they understood it. Moreover, wing warping and the canard surface in front of the wings made the Wrights' 1900 glider unique.

As the final step in their design process, the Wrights needed to size the glider. How large should the wings be? How much will the flying machine weigh? These are not unrelated questions; the answer to one affects the answer to the other. The Wrights had a design specification – simply to get into the air with a pilot on board. To satisfy this specification, the Wrights had to establish a design point around which they could size the glider. How they chose a design point, and what it was, is not explicitly made clear in their writings, but from close examination of their notes and letters, we can piece together their most likely approach. To begin with, the Wrights understood and used an existing equation for calculating lift:

[13] Lilienthal, Otto, *Der Vogelflug als Grundlage der Fliegekunst*, R. Gaertners Velagsbuchhandling, Berlin, 1889 (Translated by I.W. lsenthal and published in 1911 as *Birdflight as the Basis of Aviation*, Longmans, Green, London, 1911.

$$L = kSV^2C_L \tag{1}$$

L is lift, S is wing area, V is flight velocity, C_L is the lift coefficient, and k is Smeaton's coefficient. With L in units of pounds, S in square feet, and V in miles per hour, the classically accepted value for k, based on experiments from the eighteenth century, was 0.005. For steady, level flight, the weight W of the aircraft is exactly balanced by the lift, i.e., L = W, and the equation in the form used by the Wrights is

$$W = 0.005\,SV^2C_L \tag{2}$$

This equation per se cannot be found in the Wrights' papers, but it was in the literature they studied. Expressed in metric units, the equation first appeared in Lilienthal's book *Birdflight as the Basis of Aviation*[14] published in 1889, and then two years later in Langley's book *Experiments in Aerodynamics.*[15] Langley went on to explicitly point out that the lift coefficient, C_L, is a function of the angle of attack of the wing, and is "a function to be determined by experiment." During the design of their 1900 glider, the Wrights did not have access to Lilienthal's complete book, but they owned a copy of Langley's *Experiments in Aerodynamics* that they purchased from the Smithsonian Institution in 1899 for the price of one dollar. The Wrights made clear their understanding of Equation (2) when Wilbur wrote in his paper published by the Aeronautical Society of Great Britain in 1901 that:[16]

> Rule: The angle of incidence is fixed by area, weight, and speed alone. It varies directly as the weight, and inversely as the area and speed, though not in exact ratio.

Solving Equation (2) for C_L, and emphasizing that C_L is a function of angle of attack by denoting it as $C_L(\alpha)$, we have

$$C_L(\alpha) = \frac{W}{0.005\,S\,V^2} \tag{3}$$

Equation (3) is the mathematical equivalent of Wilbur's "rule."

How did the Wrights use Equation (3) for their design? Equation (3) has four variables; weight, wing area, flight velocity, and lift coefficient.

[14] Ibid, p. 136

[15] Langley, Samuel P., *Experiments in Aerodynamics*, Smithsonian Institution, Washington, DC, p. 60.

[16] Wright, Wilbur, "Angle of Incidence," *Aeronautical Journal*, July 1901, pp. 47–49.

Knowing any three, the fourth can be calculated from Equation (3). Also, weight and wing area are not independent; the larger the wing, the heavier the structure. It is clear that the Wrights chose a design angle of attack of three degrees, which corresponded to near the maximum ratio of lift to drag[17] from the Lilienthal table. In his paper to the Western Society of Engineers on September 18, 1901, Wilbur singled out for discussion Lilienthal's data at three degrees angle of attack,[18] and he explicitly mentions Lilienthal's value of 0.545 for the lift coefficient at this angle of attack. Clearly, the Wrights had chosen three degrees as their design angle of attack, and they used $C_L = 0.545$ in Equation (3). Wilbur goes on to discuss how he had to change the design wing area of 200 square feet of their original design because he could not find a suitable length of wing spar material on his way to Kitty Hawk. He related: "We were compelled to make it only 165 square feet in area, which, according to the Lilienthal tables, would be supported at an angle of three degrees in a wind of about 21 miles per hour." Their original design velocity was 18 mph. The only way that he could have obtained this new result for wind velocity was from Equation (3), still using Lilienthal's value of 0.545 for the lift coefficient at three degrees angle of attack.

Wilbur, in discussing part of their basic design philosophy for the 1900 glider, wrote in his September 18 paper that they wanted to be able to "practice by the hour instead of by the second." He went on to say that:[19] "It seemed feasible to do this by building a machine which could be sustained at a speed of 18 miles per hour." (Why they chose 18 mph, rather than a normal round number of 15, or 20 mph, is not clear. It might have something to do with the wind velocities that they assumed would exist at Kitty Hawk.)

Examining Equation (3), the weight and wing area form a ratio, W/S, which today is called the wing loading; it is one of the most important airplane design parameters. The Wrights knew this, and for their design point, they obtained from Equation (3), W/S = 0.9 lb/ft^2. The Wrights

[17] Based on the normal and tangential force coefficients tabulated in the Lilienthal table, the ratio of lift-to-drag for one degree angle of attack is 18.4; for two degrees, 19.47; for three degrees, 19.1, and for four degrees, 17.21. The angle of attack of two degrees yields the highest lift-to-ratio, but three degrees is very close to the highest. Moreover, if the drag due to the vertical struts and supporting wires is added, the angle of attack for maximum lift-to-drag ratio will shift to a higher value. So the Wrights' choice of three degrees as a design angle of attack, as far as they could tell, corresponded to the maximum lift-to-drag ratio.

[18] McFarland, pp. 105, 112. [19] McFarland, p. 103.

converged on a wing area of 200 square feet, which from Equation (3) they knew would support a total weight of 180 pounds. When the 1900 glider was built, the total weight including pilot was 190 pounds, close enough to their design calculations. However, upon arrival in Kitty Hawk, faced with the reality of having to construct a smaller wing due to "the impossibility of obtaining suitable material for a 200 square-foot machine," Wilbur had to recalculate the flight velocity for the smaller 165 square-foot wing, using the changed value of $W/S = 190 lb/165 ft^2 = 1.15 lb/ft^2$. From Equation (2) with $W/S = 1.15 lb/ft^2$, he found V =21 mph, as he noted in his September 18 paper.

Therefore, the design point for their 1900 glider is clear: the Wrights specified the angle of attack (hence lift coefficient) that yielded the maximum lift-to-drag ratio, and they specified the flight velocity. The required wing loading, W/S, then fell out from Equation (3). From this value of W/S, the wing area S and the weight W were obtained, taking into account that the weight depends in part on the size of the wing. Apparently, with a few refinements, the Wrights stuck with this design method through the design of their 1902 Flyer.

DESIGN REFINEMENT

The Wrights viewed the flight tests of the 1900 glider with mixed emotions. On one hand, the glider was a failure; it produced far less lift than the Wrights had calculated, requiring much higher wind velocities to sustain itself in the air. As a result, their total flying time in 1900 was reduced to about two minutes; the Wrights had expected hours. On the other hand, the glider was a success in that the wing warping, the canard configuration, and the strong structural construction worked well. In his 1901 paper to the Western Society of engineers, Wilbur noted the following about their experiments at Kitty Hawk in 1900:[20]

> Although the hours and hours of practice we had hoped to obtain finally dwindled down to about two minutes, we were very much pleased with the general results for the trip, for setting out as we did, with almost revolutionary theories on many points, and an entirely untried form of machine, we considered it quite a point to be able to return without having our pet theories completely knocked in the head by the hard logic of experience, and our own brains dashed out in the bargain.

[20] McFarland, p. 107.

Based on their flight tests in 1900, Wilbur and Orville began a series of design refinements for their next glider. They addressed the principal problem – not enough lift – by applying the same methodology they used for the design of the 1900 glider, plus some common sense. The wings of the 1900 glider did not produce enough lift: solution – make the wings of the new glider larger. They used their methodology based on Equation (3) to tell them how much larger. The Wrights appreciated the importance of wing loading, W/S, in airplane design. In his letter to Chanute on May 12, 1901, Wilbur wrote about the new glider: "The glider itself will be built on exactly the same general plan as our last year's machine but will be larger and of improved construction in its details." He went on to give details on the size of the new wing and the estimated weight. Both the Wrights and Chanute knew the importance of these two design features, and clearly the Wrights appreciated the importance of the ratio W/S as a design parameter. Wilbur wrote in his paper to the Western Society of Engineers:[21]

> When the time came to design our new machine for 1901, we decided to make it exactly like the previous machine in theory and method of operation. But as the former machine was not able to support the weight of the operator when flown as a kite, except in very high winds and at very large angles of incidence, we decided to increase its lifting power. Accordingly, the curvature of the surface was increased to 1 in 12, to conform to the shape on which the Lilienthal table was based, and to be on the safe side, we decided also to increase the area of the machine from 165 square feet to 308 square feet, although so large a machine had never before been deemed controllable.

The Wrights had used an airfoil shape with a flatter camber of 1 in 22 for the 1900 glider, thinking that this would have little effect on the design lift coefficient of 0.545 at three degrees of angle of attack obtained from the Lilienthal table. Second-guessing that a smaller camber might have actually reduced the value of the wing lift coefficient, causing some of the problem of their insufficient lift, the Wrights decided to use the same 1 in 12 camber as used by Lilienthal for his aerodynamic measurements. With this, the Wrights continued to use a design lift coefficient of 0.545 at three degrees angle of attack for their 1901 glider. This, of course, gives from Equation (3) the same design wing loading of 0.9 pounds per square foot at a design velocity of 18 mph, as before. But Wilbur had put in an extra consideration, "to be on the safe side" by increasing the wing area, thus decreasing the wing loading, W/S. The Wrights knew that an aircraft with

[21] Ibid.

a smaller wing loading could lift itself off the ground at a lower flight velocity. The 1901 glider, when built, had a wing area of 290 square feet (Wilbur's value given in his paper of 308 square feet included the 18 square-foot area of the canard surface), and weighed 240 pounds, yielding a wing loading of 0.83 pounds per square foot. With this wing loading, the flight velocity calculated from Equation (3) is 17.5 mph. This was why Wilbur noted in his Western Society paper about the 1901 glider:[22] "According to calculations it would obtain support in a wind of 17 miles per hour with an angle of incidence of only 3 degrees."

The glider that the Wrights took to Kill Devil Hills in 1901 was simply a refined and larger version of the 1900 glider. In what was to become a time-honored tradition in airplane design, the new glider contained all the essential features of the previous design that worked, but had some modifications that the Wrights hoped would correct the deficiencies of the previous design.

Orville and Wilbur arrived at Kill Devil Hills on July 12, 1901 with the material for their new glider. The 1901 glider was larger by far than any previous glider built by their predecessors. In fact, the Wrights had crossed the Rubicon with this glider; weighing 240 pounds with a wing area of 290 square feet, this glider was too large and too heavy to be controlled as a hang glider in the fashion of Otto Lilienthal. shifting his weight to move the center of gravity for balance and control. The 1901 glider had to have a mechanical means of control, which of course it had with the wing warping and the canard surface deflection. The Wrights broke new ground by deciding at the outset to use mechanical controls.

The design modifications that Orville and Wilbur made did not work. Increasing the airfoil camber to one-twelfth to match Lilienthal's camber actually made things worse; it increased the sensitivity of the center of pressure movements as the angle of attack changed, hence making the glider more difficult to control. In the field, they immediately reduced the camber to one-seventeenth, and the glider handled much better, much more like their 1900 glider. The hoped for improvement in lift also did not occur. The 1901 glider exhibited the same discouraging deficiency in lift as did the 1900 glider. Wilbur wrote in his diary on July 29, 1901: "Afternoon spent in kite tests. Found lift of the machine much less than Lilienthal tables would indicate, reaching only about 1/3 as much." Again

[22] Ibid.

unaware that they were misinterpreting the Lilienthal Tables, the Wrights made the same mistakes that were made during the design of their 1900 glider. Their efforts to correct the lift problem were doomed from the start. In spite of this situation, Wilbur achieved far more gliding time than during 1900. He was becoming an accomplished pilot. (Indeed, Wilbur had done all the free flight glides in 1900 and 1901. Orville did not fly until the following year.)

DESIGN REVOLUTION

The Wrights left for home on August 20 in a state of despair. At home, they struggled to make sense of their flying experiences and of the disappointing deficiency in lift. Earlier, they had begun to doubt the accuracy of the Lilienthal tables, and their experience that summer essentially solidified their doubt. Then the Wrights made a decision that was to be a turning point in their airplane design process. Since Lilienthal's aerodynamic data was the best available at that time, and since their two gliders designed on the basis of that data produced only about one-third the predicted amount of lift, the Wrights decided to set aside the Lilienthal tables and generate their own aerodynamic data. To this end, they first mounted a small model wing on the rim of a bicycle wheel, with a flat plate perpendicular to the wind direction at a location on the rim ninety degrees from the wing model. With this wheel mounted horizontally in front of the handle bars of a bicycle, they peddled down the street, generating an air flow over the models. When the lift on the wing equaled the drag on the plate, the two moments about the center of the wheel were equal and opposite; in that case the wheel was balanced. The results were inconclusive, except that when balance of the wheel was obtained, the wing model was at an eighteen degrees angle of attack, much higher than the Lilienthal table predicted. This reinforced the Wrights' doubts about the accuracy of the table. Next, they fabricated a small makeshift wind tunnel from an old starch box and a small fan. Again, the results were inconclusive, and they used their first wind tunnel for only one day. Next they constructed a new, more substantial wind tunnel, and had it operating by mid-October in the back of their bicycle shop. This was not the first wind tunnel in history; about a dozen other wind tunnels had been previously built and used for various purposes since the Englishman Francis Wenham invented the wind tunnel in 1870. But the Wrights built their tunnel for a single purpose – to measure the lift and drag on a number of different airfoil and wing shapes specifically to guide them in the design of their next glider. Their tunnel was utilitarian and unpretentious; it had a flow duct six feet long with a square cross section sixteen inches on each side, and the airflow was driven by a fan that pushed air into the entrance with a velocity of about 30 mph. But the instruments they designed for measuring the aerodynamic properties were innovative. Orville and Wilbur designed a balance that measured not the lift force but rather the *lift coefficient* directly. The design was ingenious; nothing like that had been built before. Moreover, they designed a second balance that measured directly the lift-to-drag ratio. From the measurements obtained from both balances, C_L from the first and $L/D = C_L/C_D$ from the second, they backed out the drag coefficient. Details on the mechanical design are nicely discussed by Jakab[23] and Wald.[24] Suffice it to say that the Wrights' wind tunnel balances are a testimonial to their design genius.

From mid-October to December 7, 1901, the Wrights tested over 200 different wing models, with different planform and airfoil shapes. These experiments produced the most definitive and practical aerodynamic data on wings and airfoils obtained to that date. (Except for bits and pieces of data shared with Chanute, the Wrights never published their wind-tunnel data. It was not until 1953 that the data was published, when McFarland included extensive tables of the Wright brothers' wind-tunnel data in his editing of the Wrights' papers.) By far, the most important design result to emerge from the Wrights' experiments was the aerodynamic benefit of high aspect ratio. Everything else being equal, the higher the aspect ratio, the higher the lift and the lower the drag. Definitive data on the benefit of high aspect ratio wings had already been published by Langley in 1899 in his *Experiments in Aerodynamics* (Smithsonian Institution, Washington, DC, 1891, p. 60), but the Wrights appeared to not have noted Langley's data, or they simply ignored it.[25] Also, at this time the Wrights finally became aware that the data in the Lilienthal tables applied to only the specific aspect ratio of 6.48. The aspect ratio of their 1900 glider was 3.5 and that of the 1901 glider was 3.3; no wonder the Wrights calculations of the lift of these gliders did not agree with the measured values. Nevertheless, from this time on, the Wrights considered the Lilienthal table moot. From their wind-tunnel data, they had their

[23] Jakab, pp. 128–138.

[24] Wald, Quentin, *The Wright Brothers as Engineers: an Appraisal and Flying with the Wright Brothers, One Man's Experience*, published by the author, 1999, pp. 26–32.

[25] Anderson, John, D., Jr., *Inventing Flight: The Wright Brothers and Their Predecessors*, Johns Hopkins University Press, Baltimore, 2004.

own tables of aerodynamic coefficients, which were much more extensive and detailed than any from the past.

Compared to their two previous gliders, the Wrights' 1902 glider reflected a revolution in its design features. Paramount was an aspect ratio of 6.7 – much larger than their previous gliders. The new wings had an area not much greater – 305 square feet compared to the 1901 glider wing area of 290 square feet. They simply made each wing longer and narrower. They also designed the airfoil camber to be quite small – one-twenty-fifth compared to the value of one-twelfth they had started with in their 1901 glider trails. (By then, the Wrights were so confident in their understanding of airfoil properties that the airfoil shape adopted for the 1902 glider was not even one specifically tested in the their wind tunnel.) Even their forward canard surface had an aspect ratio larger than the previous year.

The new design had another striking difference – a vertical tail at the rear of the machine, consisting of two side by side vertical surfaces, giving a total vertical tail area of 11.7 square feet. They intended the vertical tail to counteract a disturbing yawing motion that Wilbur had sensed in a few of his 1901 glides. He briefly mentioned the problem in an August 22, 1901 letter to Chanute: "The last week (of flights in 1901) was with very great results though we proved that our machine does not turn (i.e. circle) toward the lowest wing under all circumstances, a very unlooked for result and one which completely upsets our theories as to the causes which produce the turning to the right or left."[26] The problem was caused by the increase in wing drag due to the production of lift – called *induced drag* or *drag due to lift*. For example, to roll to the right and hence turn to the right, the wing warping decreased lift on the right wing and increased lift on the left wing. Consequently, the drag on the right wing decreased and the drag on the left wing increased, causing the airplane to yaw (pivot) towards the left, in the opposite direction of the turn. This yawing motion is precisely what Wilbur had experienced but had not expected. The brothers did not understand what made it happen, but they correctly theorized that this undesirable yawing action could be suppressed by adding a vertical tail at the rear of the machine. A vertical tail provides a "weather-vane" effect which would swing the glider back to its proper zero-yaw position if any unexpected yaw motion occurred during a turn. Hence, their new design had a vertical tail.

[26] McFarland, p. 84.

FIGURE 2.6. Wright brothers'1902 glider, flying beautifully at Kill Devil Hills. (National Air and Space Museum, SI-2004-41001.)

Their 1902 glider was a spectacular success in comparison to the previous two. Arriving at Kill Devil hills on August 28, the Wrights spent two complete months flight testing the new glider. The 1902 glider, shown in Figure 2.6, flew beautifully, and aesthetically was more graceful looking than their previous gliders. The high aspect ratio of the wings is clearly seen in this photograph.

One more fundamental design change was made in the field in order to correct a new problem introduced by the vertical tail. Whenever an airplane rolls (banks), it also experiences a sideslip (sidewise motion) in the direction of the lower wing. Because of this sidewise slipping, the entire side of the airplane that faces into the sideslip feels a component of the relative air velocity at right angles to the side. This includes the fixed vertical tail, which feels a component of force perpendicular to its surface due to the sideslip. If left uncontrolled, this can set up a spiraling motion of the airplane into the turn, that is, a corkscrewing motion around the tip of the lower wing. Today, this is called a spin. Such spins happened a few times during the Wrights' early flights in 1902, with the result of the glider falling out of the sky and the wing tip slamming into the sand. The Wrights called such an event "well-digging." On October 3, during a sleepless night, Orville conceived the solution to the problem – design the vertical tail so that it could pivot as a rudder. In this way, when the tail started to swing in an undesirable direction in a sideslip, rudder deflection controlled by the pilot would counteract and stop the undesirable yawing motion. Wilbur readily accepted the idea, and improved on it by suggesting that since the pilot already had a lot to do controlling the glider, that the rudder deflection be connected with the wing warping mechanism so that the rudder would automatically deflect in the correct direction when the wings were warped. Within a few hours that morning, meshing their ideas, the Wrights developed an effective control mechanism for yaw on their flying machine. In the process, they redesigned the vertical rudder to have only a single surface, feeling that would be just as effective as the double surface. With this, their flight control system was complete, and the 1902 glider became the first flying machine with full control around all three axes – pitch, yaw, and roll. After the movable rudder was installed, the tailspins and well-digging disappeared. The problem was solved.

Another technical contribution was the overall structural design of the glider – its lightness and durability. The Wright borthers used lightweight, flexible, and resilient wood that yielded but usually did not break whenever the glider had a hard encounter with the ground. The fabric of the wings, canard, and vertical tail played a structural role, being applied on the bias so as to add strength to the machine. Also, amazingly enough, the wooden ribs and spars were not rigidly attached to each other; the cloth covering and pockets around the joints held the structure together. This is one reason why their machines were readily repaired whenever any damage occurred.

All these contributions added up to produce an excellent *system*. In the 1902 glider, the Wrights had good aerodynamics, good flight control, and good structures, and these elements all worked together to make that glider a tremendous success. The Wrights' 1902 glider was the first truly successful *aeronautical system* – a major revolution in airplane design.

The Wrights left Kill Devil Hills for home on October 28. What a change from the year before! They now had the most successful flying machine ever developed. Moreover, Orville started flying in 1902, and they each accumulated more than an hour of total flight time in the air. Their aeronautical technology had advanced far beyond that of any other – more than Langley, more than Chanute, even more than Lilienthal. Only one system aspect was missing – propulsion. To date, gravity had been their mode of propulsion – gliding downhill. The Wrights were now ready to take the next step.

DESIGN SUCCESS

The Wrights accumulated extensive flight test data with their successful 1902 glider. In their design methodology, the flight data was more important than their wind-tunnel data. Wilbur wrote to George Spratt on February 14, 1902, even before the 1902 glider was built and flown: "Except for the time lost in waiting for suitable weather, I would prefer natural wind to any fan arrangement."[27] The major impact of their wind-tunnel tests on the design of the 1902 glider was the use of the large aspect ratio wing. The airfoil shape used for the wing was different than any tested in their wind tunnel. It is clear, however, that at some stage the Wrights intended to use their wind-tunnel data as part of their design methodology. Wilbur shared with Chanute in a letter on March 11, 1902: "My idea has been that in designing a gliding machine the experimenter wishes to know the weight that he can sustain with a given area, angle, and speed; and the angle at which he can glide."[28] Then he contrasted that with powered machines: "While in power machines it is the lift and *thrust* that is desired." Wilbur then stated the major function of the wind-tunnel data: "My plan has been to design the tables (of wind-tunnel data) so as to give information on these particular points with the least possible figuring." The Wright prepared their wind-tunnel data for convenient use in their design process.

The design methodology used for their 1903 Flyer was more extensive than that used previously because of the addition of a power plant and propellers for thrust. Nevertheless, the estimate of weight and size for the airplane most likely followed their time proven approach. Weight was most likely estimated based in part on their experience with the 1902 glider. For example, Orville cataloged a detailed weight breakdown of the 1902 glider in his notebook on October 21, 1902, noting such items as "front rudder complete 10¼ lbs. Both (wing) surfaces and uprights and wire 87 lbs. Vertical tail weighs with wires, braces, etc. 3¾ lbs. This leaves 15 lbs for skids, pulleys, wire and rope." They also knew the center of gravity location as "approximately 18 inches from the front edge." This detailed weight breakdown anticipated the need for similar detailed weight estimates for the powered machine. They knew the powered machine would be larger and heavier. On December 11, 1902, Wilbur wrote to Chanute: "It is our intention next year to build a machine much

[27] McFarland, p. 217. [28] McFarland, p. 225.

larger and about twice as heavy as our present machine. With it we will work out problems relating to starting and handling heavy weight machines, and if we find it under satisfactory control in fight, we will proceed to mount a motor."[29] He was more precise in a letter on December 29 to George Spratt: "We are thinking of building a machine next year with 500 sq. ft. surface, about 40 ft. x 6 ft. 6 inches. This will give us opportunity to work out problems connected with the management of large machines both in the air and on the ground, such as starting, &c. If all goes well the next step will be to apply a motor."[30] Most likely, the Wrights had already performed a preliminary weight and sizing analysis.

Essential to this analysis is a design lift coefficient and corresponding angle of attack for the Wright Flyer. Although they never explicitly itemized their design point, we can make an educated guess. Orville included in a letter to Spratt on June 7, 1903, a rather detailed sketch of the airfoil section, including not only the shape but also the internal structural design, and stated that the camber was 1 in 20. Among all of the different airfoil shapes and wing aspect ratios tested by the Wrights in their wind tunnel, several had a camber of 1 in 20. Wilbur singled out one in particular as having the "highest dynamic efficiency of all the surfaces shown," i.e., having the highest value of maximum lift-to-drag ratio. In the table of the Wrights' wind-tunnel data published by McFarland, this airfoil is number 12; it is a wing with a parabolic-shaped airfoil and an aspect ratio of 6. Its maximum L/D occurs in the angle of attack range between the two adjacent table entries of five degrees and seven and a half degrees. In the discussion following Wilbur's presentation of his second paper to the Society of Western Engineers in Chicago on June 24, 1903, in answer to a question from the audience, he stated: "If it were desired to fly by mechanical means horizontally through the air, the best angle would be about 5 to 7 degrees; that is, the wings should be set probably in the neighborhood of 5 to 7 degrees above the horizon."[31] Wilbur was coy with this answer. By that time the Wrights had already designed the Flyer, and they know their design angle of attack. In answering the question, Wilbur knew a lot more than he let on. Indeed, his June 1903 paper focused exclusively on the design and flight testing of their 1902 glider; Wilbur intentionally said nothing about their work on a powered machine. Keying on the data from the number 12 wing model, the Wrights almost certainly picked a design angle of attack in the range of

[29] McFarland, p. 290. [30] McFarland, p. 292. [31] McFarland, p. 331.

five to seven degrees. Since the maximum value of L/D is relatively constant in this range, they would have chosen the higher value of seven degrees because it generated a higher lift coefficient. Linearly interpolating between their measured lift coefficients for five and seven and a half degrees, the lift coefficient for seven degrees is 0.668. Something in this range was most likely their design coefficient.

To find their design flight velocity, one has to look carefully at their notes on propeller design. They made their propeller calculations for a flight velocity of 24 mph, at which they calculated that their two propellers combined would produce 90 pounds of thrust.[32] Because in level flight thrust equals drag, clearly they had already made a drag calculation at their design point. In a letter to Charles Taylor written from Kill Devil Hills on November 23, 1903, Orville stated: "We had designed our propellers to give 90 lbs. thrust at a speed of 330 rev. per minute (about 950 of engine), which we had figured would be the required amount for the machine weighing 630 lbs."[33]

The design wing loading can be obtained from Equation (1), recognizing that lift equals weight in level flight.

$$\frac{W}{S} = \frac{L}{S} = kV^2 C_L \tag{3}$$

By this time, the Wrights were using the much more accurate value of Smeaton's coefficient of 0.0033. They knew from the beginning of their work in 1900 that the more accurate value of Smeaton's coefficient was 0.0033. Samuel Langley had measured this value, and published it in his *Experiments in Aerodynamics* in 1891. The Wrights accepted this measurement; it was the only data from Langley that was directly used by the Wrights. However, the Wrights assumed, incorrectly, that Lilienthal had used the classic value of 0.005 in calculating his aerodynamic coefficients, and so, for consistency, they employed the same incorrect value of 0.005 while using the Lilienthal data in 1900 and 1901. In reality, Lilienthal did not need or use Smeaton's coefficient to calculate his aerodynamic coefficients, and this is one of the three misinterpretations made by the Wrights of Lilienthal's data. See Anderson, *A History of Aerodynamics*, Cambridge University Press, 1998, pp. 205–216, for the complete story. In any event, when the Wrights employed their own wind-tunnel data beginning in 1902, they knew to use the correct value of $k = 0.0033$ in Equation (3).

[32] McFarland, p. 606. [33] McFarland, p. 386.

From the above discussion, a likely design point used by the Wrights was V= 24 mph, C_L = 0.668, and W = 630 lbs. From Equation (3), the design wing loading is

$$\frac{W}{S} = \frac{L}{S} = kV^2C_L = (0.0033)(24)^2(0.668) = 1.27 lb/ft^2$$

For the design weight of 630 pounds, this yields a design wing area of S = W/(1.27) = 630/1.27 = 496 ft^2. This result is very close to the wing area of 500 square feet of the Wright Flyer.

We are not certain what design point the Wrights actually used. The above discussion is somewhat like Monday morning quarterbacking. We do not know if the Wrights used their wind-tunnel data to pick a design point. The airfoil shape used for the Wright Flyer was not precisely one tested in the tunnel. Indeed, by the time they designed the Flyer, they had a large amount of flight test data from the 1902 glider, including measurements of lift and drag. Moreover, by this time the Wrights were concerned about applying their wind-tunnel data obtained on small models (low Reynolds number data) directly to a full-size flying machine. Much later, in 1939, Orville shared this concern with George Lewis, head of research for the National Advisory Committee for Aeronautics, indicating that they were reluctant to rely too heavily on their data obtained with small models for application to a full-scale flying machine.[34] Indeed, they used some of their lift and drag measurements for the 1902 glider to attempt a calibration of the wind-tunnel data.

I have presented here a likely approach taken by the Wrights for sizing their aircraft, a methodology starting with the 1900 glider. It is based on picking a design lift coefficient and flight velocity, calculating the wing loading, and then obtaining both the final design weight and wing area iteratively to match the wing loading The iteration is necessary because the weight in part depends on the size of the wing structure, and area of the wing depends on the weight that it needs to lift. Other authors, including this one,[35] have in the past suggested that the weight and wing area were somehow obtained first, a design lift coefficient chosen, and then the necessary flight velocity to get off the ground calculated from Equation (1). Within the framework of the present more in-depth study of

[34] Lewis, George W., "Some Modern Methods of Research in the Problems of Flight," *Journal of the Royal Aeronautical Society*, October 1939, pp. 771–777. Also, see Jakab, Peter L., *Visions of a Flying Machine*, Smithsonian Press, Washington, 1990, p. 200.

[35] Anderson, John D., Jr., *The Airplane: A History of Its Technology*, American Institute of Aeronautics and Astronautics, Reston, Virginia, 2002, p. 118.

the Wrights' design methodology however, the approach given here seems more likely. The Wrights understood the importance of wing loading as a parameter in airplane design. Indeed, in a letter to Chanute on September 26, 1901, Wilbur even compares the fling qualities of various types of birds by comparing their wing loadings.

The design methodology used for the 1902 glider carried over to the 1903 Flyer with an added consideration – propulsion. At the design point, the engine propeller combination had to produce enough thrust to pull the Wright Flyer at its design velocity of 24 mph. In steady level flight, thrust equals drag, so the Wrights had to make a decent drag estimate for the Flyer in order to know how much thrust would be required. Then they would design the engine propeller combination to produce that amount of thrust. So the problem boiled down to an accurate drag prediction, a feat that is still problematical today. The precise details of the Wrights' drag calculation for the 1903 Flyer are not available. But we have enough inklings from their writings to piece together a reasonable picture of their methodology.[36]

Right from the beginning, with the design of their first glider in 1900, the Wrights were concerned with drag, and they made an effort to predict it in advance of their first flights. At this early stage, they considered the total drag of the glider to consist of two parts (using the terminology of that time):

$$\text{Total drag} = \text{surface drift and framing resistance}$$

where the two parts are defined as:

surface drift: drag of the wings. The Wrights used the word "drift" to denote that part of the total drag due to the wings.

framing resistance: drag of everything else – struts, wires, fuselage, frame, etc.

This nomenclature was borrowed from Chanute, who itemized the elements of resistance in his *Progress in Flying Machines*[37] as the "drift" of the wings and the "hull resistance." For the 1900 glider, the Wrights predicted framing resistance from inaccurate and unreliable empirical data existing in the literature, such as from Chanute's book and the *Aeronautical Annual* from 1897. They used Lilienthal's data for the surface drift.

[36] See Anderson, John D., Jr., "Aerodynamic Drag Reduction: Technical Evolution and Revolutions Since the Wright Flyer." National Aerospace Conference Proceedings, Wright State University, 1998, pp. 166–181.

[37] Chanute, *Progress in Flying Machines*, 1894, p. 10.

When they compared their drag prediction with actual measurements taken with the 1900 glider, they made some adjustments in their estimate of framing resistance, but continued to use the Lilienthal table for drift. Once again, their drag predictions were completely unsatisfactory.

Taking a different approach for the drag calculation for the 1902 glider, the Wrights used their own wind-tunnel data for drift, and they measured the drag coefficient of wing struts of different cross-sectional shapes, picking the lowest drag shape – a rectangular shape with rounded corners at the front and back. They included the increase in drag due to the aerodynamic interference of one wing being placed above the other in the biplane configurations; they had observed such interference drag in their wind-tunnel tests, and they called it the "loss from superposing." The drag prediction methodology for the 1902 gliders was noted by Wilbur in his letter to Octave Chanute of July2, 1903:

"We figure the total area of our uprights, rudder, skids, etc. at about 8 sq. ft. The effective area at 8/3 sq. ft. and the resistance in lbs. at 3 or 4 lbs. (10 miles an hour). The drift and tangential amount to be about 23 or 24 lbs. while the loss from superposing is about 3 or 4 lbs., Total resistance about 30 lb."[38]

By this stage the Wrights were able to make a more detailed breakdown of total drag into several components. Also, the way they measured the drag of the 1902 glider was now more refined. Instead of hooking the machine to a spring scale and flying it as a kite in the wind as they had done previously, they simply calculated the drag from the measured glide path angle. (The glide path angle, θ, is related to the lift-to-drag ratio through the trigonometric relation: $\text{Tan}\ \theta = 1/(L/D)$.)[39] In a wind velocity of 10 mph, they measured a drag of 30 pounds, confirming their predictions.

To calculate the drag for their 1903 Flyer, the Wrights most likely used the same drag-breakdown methodology that worked for their 1902 glider. They predicted 90 pounds of drag at a flight speed of 24 mph – their design point velocity. In turn, they had to design an engine propeller combination that would produce 90 pounds of thrust in order to overcome the drag. In a letter to Charles Taylor on November 23, 1903, Orville stated: "We have designed our propellers to 90 lbs. of thrust at a speed of 330 rpm (about 950 of engine), which we had figured would be required for the machine weighing 630 lbs." Much to their surprise and relief, when the engine propeller combination was tested at Kill Devil

[38] McFarland, p. 336.
[39] Anderson, John D., Jr., *Introduction to Flight*, 5th Ed., McGraw-Hill, 2005, p. 429.

Hills, a thrust of 132 pounds was measured. This increased thrust was important, because the actual weight of the Flyer had grown (with pilot) to 750 pounds –120 pounds more than their original design goals. (Weight increase during the design and manufacturing process is still endemic, even in modern airplane design.)

The detailed methodology and intellectual thinking with which the Wrights developed their propellers is a design case history by itself. The interested reader should study the detailed notes and calculations given in McFarland.[40] In terms of the evolution of their overall design methodology for the airplane, however, the Wright's propeller design is a sidebar, albeit an important one. The propeller was part of their system approach; the propellers had to work well in order for the flying machine to be successful. The fact that they did work well is a testimony to the advanced understanding that the Wrights, Wilbur in particular, brought to the aerodynamics of propellers. The efficiency of their final propeller shapes was on the order of 76 percent, much higher than the 50 percent efficiency of Langley's propellers, or others before them. Moreover, the Wrights designed and built their own gasoline-powered engine, which provided a marginal power output for their purposes. But combined with their highly efficient propeller, the engine–propeller combination, when tested on November 21, 1903, produced 123 pounds of thrust, prompting Orville to write in his diary that "our confidence in the success of the machine is now greater than ever before" and Wilbur to state in a letter to George Spratt on December 2 that "we will have ample propulsion."

And indeed they did. In fact, as shown in Figure 2.1, on December 17, 1903, the Wrights proved that all their systems were "ample" when the Wright Flyer, with Orville aboard, achieved the first successful heavier-than-air, controlled piloted, sustained flight in history, the ultimate demonstration of their design methodology.

In summary, Orville and Wilbur Wright were the first to develop a rational methodology for the design of an airplane – a methodology that *worked.* Wisps of their methodology can even be seen today in our modern methodology for conceptual airplane design. Figure 2.7 is taken from Anderson, *Aircraft Performance and Design.*[41] It shows the author's "seven intellectual pivot points for conceptual design." Let us compare this diagram with the Wrights' methodology. Starting at the top, the Wrights' design was driven by one essential requirement, namely, that the airplane should simply

[40] McFarland, pp. 594–640.
[41] Anderson, John D., Jr., *Aircraft Performance and Design*, McGraw-Hill, 1999.

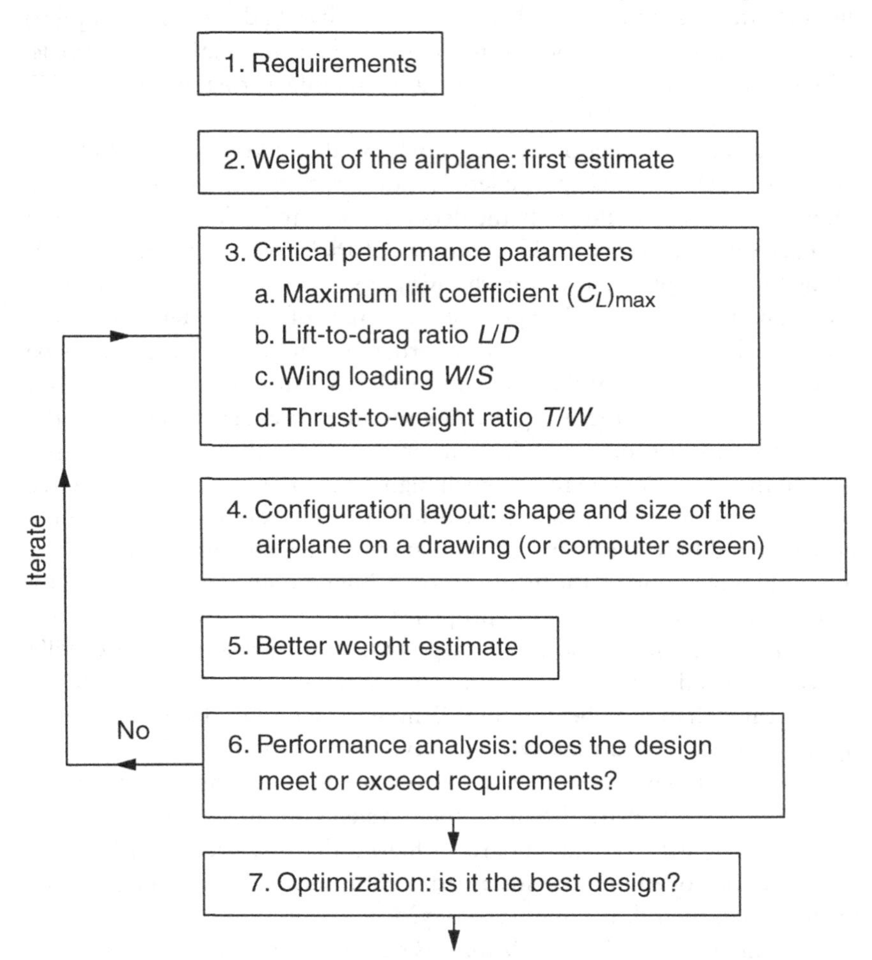

FIGURE 2.7. Seven intellectual pivot points for conceptual design.

be capable of a sustained, powered, piloted, and controlled flight. The Wright Flyer achieved this. Second, the Wrights were well aware of the importance of weight estimates, and how weight played into the critical performance factors (box 3 in Figure 2.7). Of these performance parameters, they actually used the lift-to-drag ratio and the wing loading directly in their calculations; they understood how these two parameters affected the performance of the airplane. The Wrights drew only one configuration layout (box 4) of their machine during their design process, a crude three-view

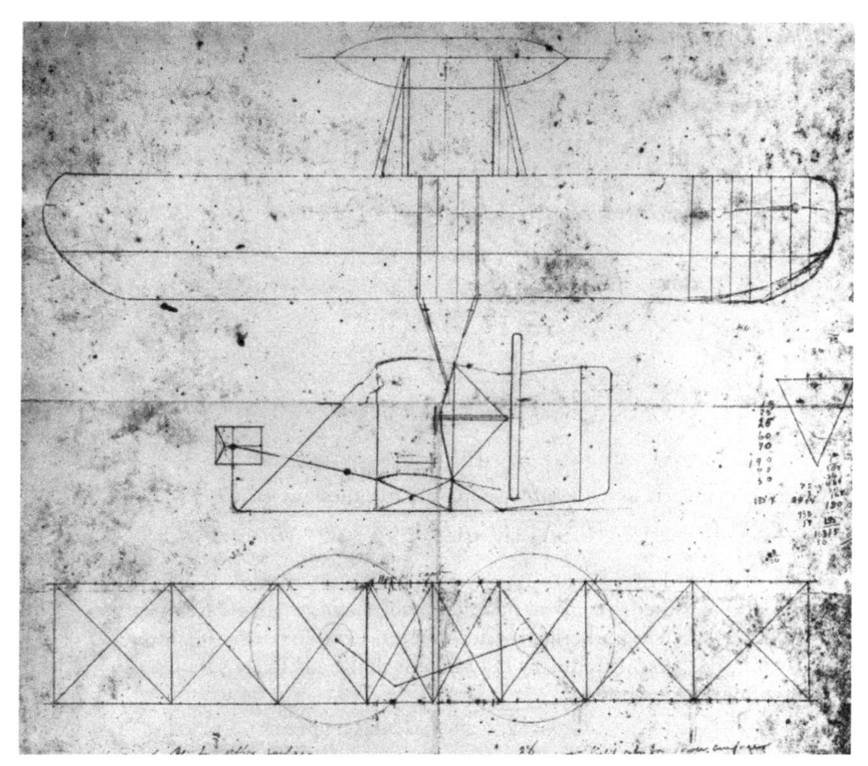

FIGURE 2.8. Wright brothers' preliminary three-view sketch of their 1903 machine, drawn in pencil on brown wrapping paper.
(Library of Congress.)

drawing of the Wright flyer on brown paper. This three-view, now in the collection of the Franklin Institute in Philadelphia, is shown in Figure 2.8.

Returning to the modern conceptual design process shown in Figure 2.7, their methodology went no further than box 4; formal optimization techniques lay far into the future. Of most importance in their methodology, however, was that they intelligently chose a design point on which to hang the details of their design – such as a design angle of attack (hence design lift coefficient) and a design velocity. They were not flailing in the dark; rather, their design point provided focus throughout the process.

For all these reasons, Orville and Wilbur Wright were the first of the *grand designers*.

3

Setting the Gold Standard

The Design Methodology of Frank Barnwell

There is evidently considerable scope yet for guess-work.

F. W. Lancaster, *The Wright and Voisin Types of Flying Machines*

It is to be noted that (Frank Barnwell's) general method of design is approved by other aeroplane designers who have been successful in producing efficient and effective aeroplanes. Consequently the new arrival in the aircraft industry may take it that he is fairly safe in following that method.

C. G. Grey, Editor, *The Aeroplane*

On September 9, 1916, the newly designed Bristol F2A fighter took off for the first time from Filton, England, with Captain C. A. Hooper at the controls. Shown in Figure 3.1, the F.2A was a modern, somewhat conventional configuration for the middle of World War I. In a period when the first flight of a new aircraft design was somewhat problematical and fraught with danger, the F.2A lifted off the ground with no trouble. Captain Hooper reported, however, that the new airplane could not climb above 6,000 feet. The problem was the altimeter, and after proper adjustment the airplane climbed easily to 10,000 feet in the first fifteen minutes on a subsequent flight. Indeed, the official trials were so successful that in November, 1916, an order was placed for fifty aircraft and Field Marshal Sir Douglas Haig demanded that two squadrons be ready for the 1917 spring offensive in France. The F.2A, after some teething pains caused by pilots inexperienced in employing the proper tactics for the new fighter, proved extremely successful. "They tore into the enemy, with most effective and immediate result," wrote Major

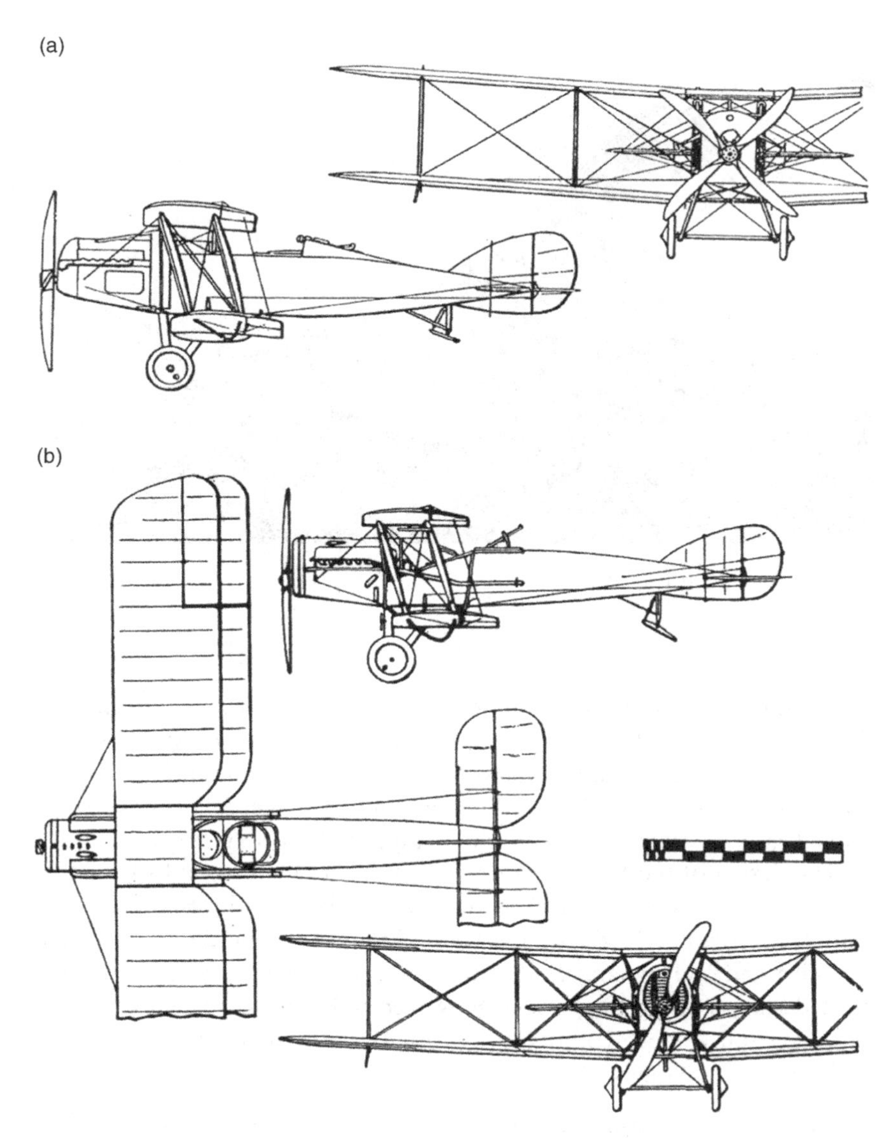

FIGURE 3.1. (a) A two-view of the Bristol F.2A fighter.
(b) A three-view of the Bristol F.2B fighter.
(c) The Bristol F.2A fighter.
(National Air and Space Museum, SI-97-17255.)

(c)

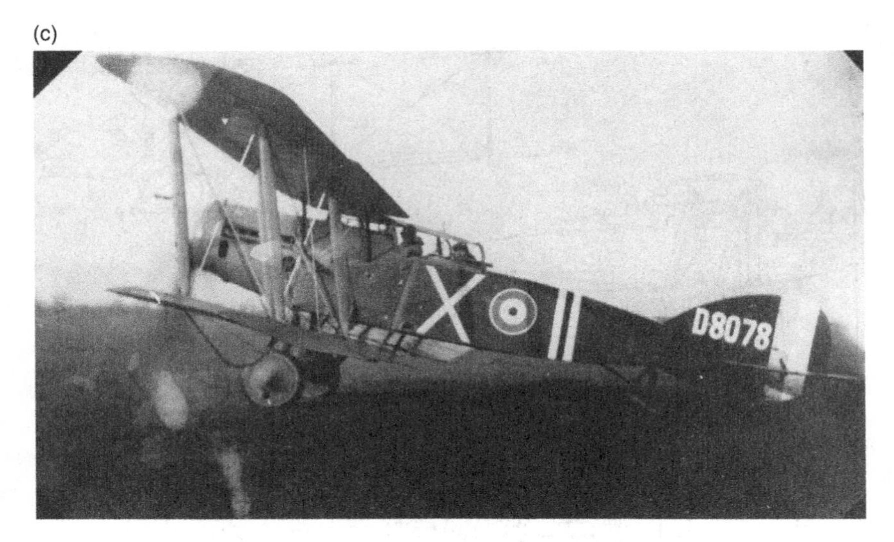

FIGURE 3.1. (*cont.*)

G. P. Bulman.[1] Those results supported the original hopes for the Bristol Fighter, and the order was increased to 602 machines, most of them a modified version called the Bristol F.2B, also shown in Figure 3.1. Ultimately, over 5,500 Bristol Fighters were produced. They were so valued that the last F.2B was not taken out of RAF service until 1932.

The design of the F.2A began in July 1916 and two months later the first airplane was in the air. The aircraft was designed by Frank Sowter Barnwell, Bristol's chief designer. He was not a household name such as Sopwith and DeHaviland at that time, and is virtually unknown in the United States even to the present. However, in Frank Barnwell's obituary published on August 10, 1938, the noted editor of the periodical *The Airplane*, C. G. Grey wrote:[2]

> Frank Barnwell was beyond question one of the best aeroplane designers in this country or in the World. No other designer has turned out so many first-class aeroplanes which have become historic.

[1] Bulman, G. P., "Frank Barnwell," *Flight*, Vol. 65, March 19, 1954, pp. 338–342; see also the *Aeronautical Journal*, Vol. 58, June 1954, pp. 382–395.

[2] *The Aeroplane*, August 10, 1938, p. 162.

Frank Barnwell is one of the Grand Designers. The basic intellectual skeleton of the airplane design methodology used today originated with Barnwell in 1916. In this chapter, we will see that in an almost natural fashion, Barnwell's approach to conceptual airplane design closely followed the intellectual pivot points outlined in Figure 2.7. He started with the requirements for the new airplane, looking at what its main purpose was and what it was supposed to do. He then made an approximate weight estimate because weight plays strongly in the next step, the estimated performance of the new design. With the aid of a configuration layout – a drawing that helps to establish the shape and size of the new airplane – he made a better, more precise weight estimate, which resulted in a better, more precise performance analysis. If he felt it necessary, he repeated this process in a somewhat iterative fashion until he was satisfied with the conceptual design. Moreover, he wrote all this down in a book for all to see. Barnwell's methodology is a common sense approach, but when his book appeared in 1915, it hit the airplane design community like a blockbuster. Unlike the usual evolutionary model of technological advancement, Barnwell's design method was new. It was a product of his natural genius, and his innate feeling of how to go about conceptualizing a new airplane. This chapter tells his story and explores his design methodology.

BARNWELL: THE MAN

From 1976 to 2012, on the wall of the old Beyond the Limits Gallery at the National Air and Space Museum in Washington, DC, there was a preliminary sketch of the Bristol fighter drawn by Frank Barnwell in 1916. The hand lettering on the drawing was immaculate. Barnwell's papers are now in the library of the Royal Aeronautical Society. The logic and organization of his entries in his engineering notebooks is clear and concise, and his hand lettering in these notebooks is also immaculate. Indeed, his notes are so thorough, so clear, so neat, as if he meant for them to be read by me almost a century later. Frank Barnwell (Figure 3.2) was exceptionally fastidious in his work and in his appearance, and his colleagues respected him all the more for it. All this was the outward appearance of the inwardly organized and focused state of mind of the airplane designer who set the gold standard for the methodology of conceptual airplane design.

Frank Sowter Barnwell was born on November 23, 1880 in Lewisham, Kent, south east of London, but grew up in Glasgow. He graduated from Feltes College Public School in Edinburgh (the Eton of Scotland) at the

FIGURE 3.2. Frank Barnwell (1880–1938).
(Royal Aeronautical Society.)

age of 18, after which he served a six-year apprenticeship with the Fairfield Shipbuilding Company at Govan, where his father was managing director. During this time, Barnwell took evening classes at the Technical College and attended Glasgow University during the winter months. He graduated from Glasgow in 1905 with an Engineering degree in Naval Architecture. (The affinity of early technical design of airplanes to a background in naval architecture was strong during the first decades of the twentieth century. For example, the first courses in aeronautics at MIT were offered through the naval architecture program. The similarities in fluid dynamics and structural design between these two disciplines made existing naval architecture programs fertile soil for the growth of new courses in aeronautics.) For reasons unknown to this author, Barnwell then traveled to the United States where he spent a year as a hull-draftsman in a shipbuilding firm.

Returning to England in 1907, he started his own firm with his brother Harold called the Grampian Engineering and Motor Co., Ltd. in Stirling. This was the beginning of Barnwell's career as an airplane designer. After several false starts with two homemade gliders and a machine powered by a Peugeot motor cycle engine, the brothers had their first halting success with a parasol monoplane powered by a two-cylinder air-cooled engine designed by Harold.

In 1910, the brothers designed and flew a tractor monoplane with a 40 horsepower twin-cylinder water-cooled engine, also designed by Harold. This aircraft won a monetary fifty pound prize from the Scottish Aeronautical Society.

The brothers' collaboration at Grampian ended in 1910 when Frank became engaged to the daughter of an Army Colonel in Stirling, and for financial reasons returned to more secure work in naval architecture. Harold remained in aviation, however, joining the Vickers Company as a test pilot at Brooklands, by that time a very popular and huge car racing track with an expansive inner field suitable for testing airplanes. But their general collaboration in aeronautics continued. Indeed, when the first flight of the Bristol F.2A, described at the beginning of this chapter, seemed to indicate a difficulty of the aircraft climbing beyond a 6,000-foot altitude, Harold was called in by Frank and was helpful in assessing the spurious result as an altimeter problem. Shortly after this, on August 25, 1917, while suffering from influenza, Harold collapsed in the cockpit while flying an experimental airplane near Brooklands; the airplane went into a spin and he died instantly in the resulting crash.

Frank Barnwell could not stay away from the airplane business. In March 1911, he took a position as a draftsman with the British and Colonial Airplane Company located in Bristol. The company had been founded a year earlier by Sir George White, a strong player in the transport industry. Sir George became convinced of the great potential of the airplane as a means for transport after watching flights at Cannes in 1909, and his new company quickly became recognized for quality design and the effectiveness of its flying schools. Much later, in 1919, the British and Colonial Airplane Company was renamed the Bristol Aeroplane Co., Ltd.

When Barnwell joined the British and Colonial Airplane Company, the design work was shared among six men, one of whom was Henri Coanda, son of the Rumanian War Minister. (Coanda was later to become somewhat famous for his discovery of the "Coanda effect" – the tendency for the low speed flow of a liquid or gas to remain attached to, and to close behind the surface of a body.) In 1912, Coanda became chief designer, and Barnwell was slowly absorbed into the main design activity of the small company. In 1914, Coanda gave Barnwell his first major responsibility, namely the design of a new airplane around the fuselage of one of Coanda's earlier concepts. The result was the Bristol Scout, the first famous airplane to come from the Bristol factory. The Scout was a biplane powered by an 80 horsepower Gnome rotary engine.

With the pilot and enough fuel for three hours of flight, its weight was 950 pounds (only about two-hundred pounds heavier than the 1903 Wright flyer). Two Scouts were delivered to the Royal Aircraft Factory at Farnborough in August, 1914, and production of the airplane commenced. It served as a scout airplane in France and the Middle East as late as 1917, by which time its top speed of 97 mph made it somewhat obsolete. But Barnwell's star as an excellent airplane designer was starting to rise; we note that Barnwell recorded the necessary design and stressing calculations for the Scout in a penny notebook in that same immaculate hand lettering that was to become his hallmark.

Barnwell's early career with Bristol was interrupted in late 1914 by two related phenomena associated with the beginning of World War I. First, the War Office designated the Royal Aircraft Factory (RAF) at Farnborough as the sole designer for new aircraft, relegating the industry only to the production of these designs. The entire Filton works of Bristol was charged with the production of the RAF designed B.E.2c. The design staff at Bristol was left with no creative role. Secondly, the patriotic fervor of the time strongly encouraged young men to join the military. Moved by this push-pull environment in which Barnwell found himself, he joined the Royal Flying Corp (R.F.C.) in November 1914, and at the age of 35 he received his wings in March 1915. He continued flying for the rest of his life, up to the instant of the airplane crash that took his life in 1939.

The War Office changed its policy in 1915 in the face of terrible losses of British aircraft and pilots. Hounded by criticism from Parliament and chastised by the First Lord of the Admiralty, Winston Churchill, the War Office now supported designs from industry along with those from the Royal Aircraft Factory. One effect was the release of Captain Frank Barnwell from the R.F.C. "on indefinite leave without pay" to rejoin Bristol. Coanda had left Bristol at the beginning of the War; Frank Barnwell found himself immediately appointed as chief designer.

Barnwell now came into his own. He designed the Bristol F.2A and its almost immediate variant, the F.2B; the success of this airplane is noted at the beginning of the chapter. The airplane, named the "Brisfit" by the R.F.C., solidified Barnwell's reputation as an airplane designer, a reputation that continued to grow over the next two decades.

There was, however, one other interruption of Barnwell's career with Bristol. After the Armistice in 1918, the bottom dropped out of the aircraft industry. The huge surplus of World War I aircraft effectively satisfied the market for airplanes for almost a decade after the war. The aircraft industry, without meaningful new orders, suffered greatly. New airplane designs found no market. Mired in this situation, Barnwell looked for a way out. What happened next is aptly summarized by Barnwell's close colleague and friend, Major G. P. Bulman in his 1954 article in *Flight*:

> Barnwell was clearly unhappy. There was a lack of that complete occupation which he had had throughout the war and of the satisfaction of seeing his design effort materialize in useful production. And so, in October 1921, he resigned and took a technical commission as a major in the Royal Australian Air Force, to join an experimental section for aircraft design and development in the Commonwealth. But for reasons not of his making, which some of us well understood at the time, his zest for creative activity did not receive the expected outlet, and the lamp of achievement was not rekindled. In October 1923, just two years later, he turned up again unheralded at Bristol and was re-appointed Chief Designer.

For the rest of his life, Barnwell remained at Filton, rising to chief engineer of Bristol Aircraft in 1936.

Over the course of twenty-six years, Frank Barnwell designed and laid out more than 150 different aircraft. Some were design studies and never went into production. He played the major role in all of them. The Bristol Bulldog, designed by Barnwell, became the mainstay of RAF fighter squadrons in the midyears between the two world wars. Among his last designs was the Bristol Blenheim, the famous twin-engine bomber that served in the RAF during the early days of World War II. Little known is Barnwell's penchant for propeller design. Some of his colleagues joked that he designed a propeller a day. It started as early as 1909 when Barnwell designed propellers for the flying machines designed and built by the Grampian Engineering and Motor Company. In 1927, he received the Taylor Medal for the best paper presented to the Royal Aeronautical Society; its title was "Some Notes on Airscrew Design." His handwritten notes, now in the Library of the Royal Aeronautical Society, are peppered with technical discussions and new designs for propellers.

Through all this, Barnwell continued to fly. In 1937, with some reluctance but with the interests of all involved, Bristol Aircraft convinced Barnwell, at the age of 57, to stop flying solo in any Bristol airplanes. He continued to fly privately on his own. At that time the Civil Air Guard (C.A.G.) wished to promote public interest in flying. Responding to this interest in 1938, Barnwell, in his spare time, designed for C.A.G. use a small single-seat monoplane of a total weight of 750 pounds powered by a 25 horsepower Scott Squirrel two-stroke engine. Barnwell paid for the construction of this aircraft by a syndicate of aircraft workers at Bristol

FIGURE 3.3. Frank Barnwell's grave, Alveston, England.
(image given to the author courtesy of Dr. R. J. Hill, Rolls-Royce retired, with kind permission from St. Helen Churchyard, Alveston.)

Airport (Whitechurch). The first test flight with Barnwell at the controls went well. During its second flight, on August 2, 1938, as the airplane began the first turn after takeoff, it stalled, went into a spin, and crashed. Frank Barnwell was killed instantly, shades of his brother Harold's death in 1917. Barnwell is buried, along with his wife Marjorie and his three sons – all of whom were killed while flying for the Royal Air Force in 1940 and 1941 – in the quiet churchyard of Alveston, Bristol, his home for almost twenty years (Figure 3.3).

Airplane design is a subspecialty of the more general field of aeronautical engineering. Frank Barnwell was a consummate airplane designer, but he also functioned in a broader sense as an aeronautical engineer. For example, Barnwell was an early user and proponent of wind-tunnel testing. In his notebook dated February 6, 1914, in which he was making various types of parametric calculations on the weight and balance of aircraft, he tabulates and uses wind-tunnel results obtained at the National Physical Laboratory (NPL) on a biplane, "hydro aeroplane No. 2." In a yet earlier notebook from 1913, he looked at propeller characteristics using airfoil section data obtained in an NPL wind tunnel. This is the earliest record of Barnwell's propensity for designing his own propellers. His early use of wind-tunnel data indicated a certain degree of confidence in such data at a time when almost all other aeronautical engineers, especially airplane designers, viewed wind-tunnel data as unreliable at best and totally inapplicable at worst due to the large scale effects inherent between small-scale testing at low speeds in wind tunnels and real conditions existing in full-scale flight. (As mentioned in Chapter 2, even the Wright brothers viewed the quantitative data from their 1901-1902 wind-tunnel tests as not directly applicable to their full-scale flying machine; the most important and useful result from their wind-tunnel data was their qualitative finding of the aerodynamic benefit of an increased aspect ratio.) No better example of Barnwell's support of the importance of wind tunnels is that illustrated in the February 3, 1919, meeting of the Directors of the Bristol Aeroplane Company to discuss future design policy in the wake of the collapse of new airplane orders following the end of World War I. In this meeting both Barnwell and Sir Stanley White, son of the company founder the late Sir George White, emphasized that "they wished to proceed as fast as possible with new machines which would be in advance of anything yet produced and that out-of-date designs should be discarded as soon as possible."[3] In this meeting Barnwell was adamant that a company wind tunnel was necessary to achieve this goal. As a result, Bristol Wind Tunnel No. 1 was completed in June, 1919. This was the first wind tunnel built and operated by a private airplane company in Britain. It was in continuous operation until it was destroyed by German bombs in 1942.

Although somewhat conservative (a trait of many successful airplane designers), Barnwell was not beyond supporting unusual and innovative ideas. In the Barnwell papers at the Royal Aeronautical Society there is an early memo generated by the British and Colonial Aeroplane Company entitled "Test of Floats with Object of Collecting Data for Pneumatic Wing." It is stamped and initialed "Works Manager" and dated December 19, 1911, the same year that Barnwell joined the company as chief draftsman. It deals with the idea of changing the shape and size of a flexible airplane wing by pneumatic means in order to obtain a wider range of performance. It contains the statement:

[3] Barnes, C. H., *Bristol Aircraft Since 1910*, Putnam, p. 29.

Thus the pneumatic construction solves all problems including the variable speed problems, as the wings could either be pulled in simultaneously and symmetrically to increase the speed, or acted on independently with the same movement for maintaining lateral stability, which is at present effected on the type 58 machine by wing warping.

The memo goes on to address the existing problem that affected all airplane manufacturers worldwide at that time, namely the Wright brother's legal claims to the concept and means of lateral control.

With reference to the latter, as the Wright brothers seem to hold a firm patent on same and as litigation is menacing us, there seems to be no doubt that we should not lose an instant in getting to know all there is to find out about the pneumatic wing, and the most practical way to do this is by mounting a pair of experimental wings on a type 50 monoplane at once.

I could find no further reference to the pneumatic wing in the papers. The same box of notes contains a copy of a French patent for pneumatic wings granted to Jules Houdry, Seine-et-Oise, on August 23, 1911. Perhaps that is why.

During the era of the strut-and-wire biplane,[4] from the Wright Flyer to the late 1920s, most airplanes were constructed from wood and fabric – the vegetable airplane. A notable exception was the work of Hugo Junkers who in Germany designed and built the first successful all-metal airplane, the Junkers J.1 in 1915. The J.1, made of iron, successfully completed a number of test flights over a period extending well into 1916. Junkers never turned back. Switching to a special form of aluminum called *Duralumin*, Junkers designed and constructed a continuous series of all-metal airplanes through the next two decades while most other designers clung to the vegetable airplane. Indeed, the disdain with which Junkers' designs were held in Britain is seen in a sarcastic comment by the noted English airplane designer Handley Page after hearing a learned paper by Junkers given at the Royal Aeronautical Society in 1923. Handley Page shared with the audience that the first Junkers machine that he saw was one that had crashed in landing, the fuselage having broken in a rather unfortunate place. The name Junkers had been painted on its sides, and the break had occurred just behind the k, so that when he walked around the machine to attempt to find out what it was, he read the letters "Junk." The audience was considerably

[4] John D. Anderson, Jr., *The Airplane: A History of Its Technology*, American Institute of Aeronautics and Astronautics, Reston, Virginia, 2002, Chapter 5.

amused. However, engineers at Bristol started to think about the all-metal airplane. In Barnwell's papers at the Royal Aeronautical Society, there are a number of pages in the British and Colonial Aeroplane Rough Calculation Notebooks in 1916 making weight and center of gravity calculations for an all-metal airplane. What interest there was in Britain in the all-metal airplane was not due to any inherent appreciation of the eventual technical superiority of metal construction, but rather by the growing shortage of wood suitable for aircraft manufacture. By 1916 an inferior grade of spruce began to be substituted for the preferred Grade A silver spruce due to stocks being depleted by the huge number of airplanes being constructed for the war. Drawing on his earlier training at the Fairfield shipyard, Frank Barnwell designed the M.R.1, an all-metal biplane. With its development slowed by the heavy emphasis at Bristol to mass-produce the Bristol F.2B fighter for the war effort, the M.R.1 was finally ready for testing in late 1917. Two were built and successfully flown. After the Armistice, Frank Barnwell frequently flew the M.R.1, and finally the Royal Aeronautical Establishment at Farnborough accepted it for delivery. On April 19, 1919, Barnwell personally flew the airplane to Farnborough. On landing he hit a pine tree and crashed on the airfield. Barnwell, though considerably shaken, was unhurt, but the airplane sustained major damage and no efforts were made to repair it. This incident only added to Barnwell's reputation as an erratic flyer. The airplane, however, was a technical success. Nevertheless, Bristol continued, along with the rest of the British airplane companies, to produce vegetable airplanes for another decade.

During World War I, a revolution in airfoil design occurred.[5] Conventional wisdom favored a thin airfoil section. Thin wing sections were started with Leonardo da Vinci and were carried through the nineteenth century by such pioneers as George Cayley and Otto Lilienthal, perhaps following nature's design in bird's wings. At the turn of the twentieth century, embryonic wind tunnel tests, including those of the Wright brothers, indicated that thinner airfoils had much less drag than thick airfoils. Today we know this result to be an artifact of the small wind-tunnel models tested at the low speeds created in the early wind tunnels. In technical terms, these tests were carried out at low Reynolds numbers. When used on real airplanes, however, which were much larger and flew at higher speeds, the Reynolds number was much larger, and the actual

[5] Ibid. pp. 141–150.

airfoil performance was much different than measured in early wind tunnels. Indeed, the thin airfoils used on most airplanes until 1918, when inclined to larger angle of attack, experienced premature flow separation, hence stalling earlier with consequently smaller lift coefficients than thick airfoils. Of course, nobody knew that at the time.

Nobody, that is, except Hugo Junkers in Germany. For the design of his all-metal J.1 in 1915, Junker needed to use a thick airfoil section in order to have a thick spar inside the wing, which was cantilevered to the fuselage. To this end, he built a new wind tunnel at Aachen University, and later another tunnel at his firm Hugo Junkers and Company at Dessau, specifically to study the lift, drag, and center of pressure on both thin and thick airfoils. Junkers was elated with the test results. "My most extravagant expectations were surpassed," he wrote. "The thick airfoils proved not only equivalent to the thin ones of some series, but even superior, within certain limits." With a feeling of relief, Junkers went on to design his all-metal airplanes with thick airfoil sections. Keying on these results, Reinhold Platz, Anthony Fokker's chief airplane designer, used thick airfoils with a deeper spar for the famous Dr.1 triplane and the Fokker D.VII. Indeed, the resulting enhanced aerodynamic performance and structural strength of the wing of the D.VII made it the most potent of all fighter airplanes during World War I.

In spite of this revolution in airfoil design, the virtues of thick airfoils were slow to dawn on English and American aeronautical engineers. Frank Barnwell was no exception. In a paper written for the Royal Aeronautical Society in 1919, Barnwell states:[6]

> Generally speaking, the thicker the section the lighter will be the main spars for the same strength, but the worse the optimum lift over drag, and the worse the lift over drag at small values of lift. Moreover, in a thick section the maximum lift value is not much higher than a thinner one and the critical angle tends to occur earlier.

He illustrates this discussion with an airfoil shape, not identified as a standard RAF section, with a thickness of 7.5 percent. I suspect that this shape is a custom designed "Barnwell section." Here is a case of Barnwell not always being right. But he was certainly not alone. The widespread use of thick airfoils was slow in coming. Most biplanes designed in the 1920s were the standard strut-and-wire configuration using wings with

[6] Barnwell, F. S., "Some Points on Aeroplane Design." *The Aeronautical Journal*, No. 102, Vol. 23, June 1919, pp. 301–325.

thin airfoils – a testimonial to the prevailing conservative design philosophy of building a new airplane with only small departures from the previous airplane.

Barnwell also showed conservatism in regard to one of his pet activities – the design of propellers. Through World War I, airplane propellers were single piece designs made of wood. The blades could not individually be detached, and their pitch angle could not be changed. It was well recognized at that time, however, that such fixed-pitch propellers provided maximum efficiency at only one flight velocity, and for all other speeds the propeller was operating at substandard efficiency. Clearly, a propeller with an adjustable pitch angle during flight was needed. Mechanically and structurally, this was an exceptionally difficult design problem. Moreover, the use of wood for such variable-pitch propellers was not feasible; only metal propellers would do. This accelerated the development of metal propellers, and even though the practical development of the variable-pitch propeller would not come to fruition until the early 1930s, the use of metal propellers, mainly made from Duralumin, took hold in the middle 1920s. Indeed, Junkers was already using them for his metal airplanes. In his seminal paper on propeller development,[7] the noted Curator of Propulsion at the National Air and Space Museum, Dr. Jeremy Kinney states: "A major milestone in propeller design and construction was achieved by 1925: the introduction of a standardized metal ground-adjustable pitch propeller." The US Navy quickly wrote a contract with the Standard Steel Company for a hundred of these propellers for use on their new Martin T3M torpedo bombers. The metal propeller was *in*. But with Frank Barnwell, this was not so. His favoritism for wooden propellers is shown in his notes now at the Royal Aeronautical Society. For example, his design notebook "Airfoils and Miscellaneous" dated February 6, 1925, is full of propeller calculations and comparisons. Here he compares the performance of both wood and Duralumin propellers, and concludes that "Dural-blades apparently absorb only .84 of power of wood blades." Only towards the end of his life, for the design of the Bristol Blenheim twin-engine bomber, would he turn to the variable-pitch metal propeller.

This is not to say, however, that Barnwell was always a conservative airplane designer. For example, flying in the face of extreme government opposition to monoplane in contrast to biplane aircraft, he designed the

[7] Kinney, J. R., "Frank W. Caldwell and Variable-Pitch Propeller Development, 1918–1938," *Journal of Aircraft*, Vol. 38, September–October 2001, pp. 967–976.

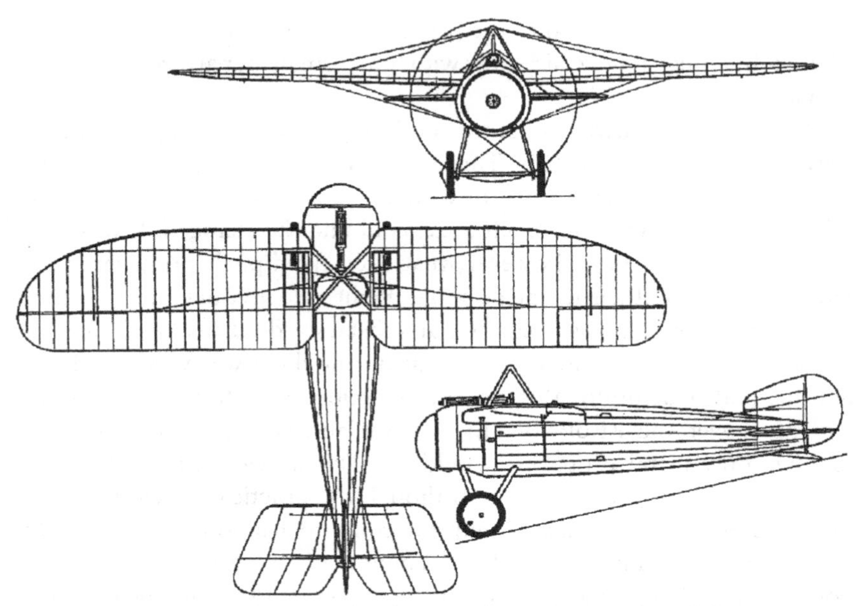

FIGURE 3.4. Diagram for the Bristol M.1C monoplane fighter.

most innovative and aerodynamically advanced aircraft of the World War I period, the small Bristol M.1C monoplane fighter (Figure 3.4). Exceptionally streamlined with a hemispherical spinner, this airplane could reach an exceptionally fast maximum speed of 132 mph. The Company expected to receive a massive production order for this high-performing fighter. Pilots of the Royal Flying Corp who had test-flown the M.1C were euphoric. However, in its infinite wisdom, the government considered the landing speed of 49 mph too high for operational aerodromes on the Western Front, and a production order was issued for only 125 aircraft. Frank Barnwell himself flew the M.1C during the test trials, and had no difficulty landing the airplane in spite of his reputation of being a somewhat erratic pilot. To no avail. Ultimately, the small number of M.1C's were relegated to serving in the Middle East. But the M.1C was proof that Frank Barnwell was indeed an innovative airplane designer.

At the time of his death, Barnwell was still in his prime as an airplane designer, and by this stage he had become somewhat of a visionary. The habitat of the airplane is the atmosphere and the nature of the airplane is to fly faster and higher. Barnwell subscribed to this philosophy. For example, in a futuristic paper to the Royal Aeronautical Society in 1937 entitled "Some Notes on Aircraft Possibilities,"[8] he blatantly stated that "speed is the outstanding superiority of aircraft over other forms of transport." He then proceeded to give detailed calculations that underscored the critical parameters for attaining high speeds. Letting W be the weight of the airplane, S be the wing planform area, and P be the power from the engine, in his own words, he states that his results "indicate the advantages of high wing loading (W/S), of low power loading (W/P), and of high altitude in obtaining high speed." He made calculations for an ideal flying wing, and then noted the major compressibility constraint on flying faster than sound. About these constraints, he said:

> As a *reductio ab absurdum*, one might note that if this perfect flying wing with its ideal power plant were 100 pounds per square foot wing loading and 5 pounds per brake horsepower power loading, its speed at 50,000 ft. ought to be 1,420 mph.
>
> Again, 1,420 mph is about twice the speed of sound in air; at speeds over about 500 mph, compressibility effects are incurred, tending to increase drag and to decrease lift; so the data, giving this 1420 mph, have become quite inaccurate.

What an understatement – a characteristic Barnwell trait. And being a detailed and precise person, he fills his paper with detailed calculations in thirteen appendices.

The next month, Barnwell published a companion paper on how to obtain high altitudes.[9] Again, full of detailed calculations, his paper contains the following conclusion:

> Finally, what might one expect to achieve in the near future, assuming the desire to go 'all out' for altitude record and assuming that it had been decided that the expenditure of energy and of money was justified? If the same engine were specified as that for a single-engined machine, I think one should go for a twin-engined monoplane with opposite turning engines and airscrews. Retractable undercarriages would now pay, as one has the nacelles into which to retract them. Metal airscrews would probably pay as they could be made reasonably light because they have to absorb very little power for their size, and have to be run 'all out' for limited periods only.

Retractable landing gear and metal propellers indeed! Frank Barnwell had finally become part of the "design revolution" that dominated airplane design in the 1930s, as we will show later.

[8] Barnwell, F. S., "Some Notes on Aircraft Possibilities," *Aeronautical Journal*, Vol. 40, March 1937, pp. 198–231.

[9] Barnwell, F. S., "Notes on the Design of Aeroplanes for Attaining High Altitude," *Aeronautical Journal*, Vol. 41, April 1937, pp. 306–323.

His close friend and colleague Major G. P. Bulman said about Barnwell in his presentation to the Royal Aeronautical Society on the occasion of the first annual Barnwell lecture initiated on March 4, 1954:[1]

> I first met Frank early in 1916, and felt always the better for, and inspired by, every meeting throughout the years which followed. He was slim of build and not very tall, with fair hair always kept short, with sparkling light-blue eyes which were often quizzical, never flinched, and could freeze. He was meticulously neat and almost fastidious in all his work; and his appearance always *soigné*, looking usually, I used to think, as if he had just emerged from a Turkish bath and was off to a point-to-point. There was nothing 'scruffy' about him. He had a rather staccato – almost hesitant – speech, and would sit long silent in conference. Then, sometimes with devastating effect, he would inject a few words which penetrated the heart of the problem. He had a curious habit of shaking the lobe of his ear as he talked, perhaps to conceal mental irritation, or as if to stimulate his brain.

About Barnwell's interaction with his associates, Bulman wrote:

> I know for myself – 12 years his junior – how ready he always was with patience and courtesy to talk and argue with younger men; and also how patiently he coped with bureaucracy – although I am told that he said once, *sotto voce*: 'We struggle against a sea of bloody fools.' He could be stubborn indeed, but there may be some who will say that, if he had failed to get acceptance of his argument, or if he had been prevented from doing what he was convinced was right, he would break off the battle and retire to his ivory tower instead of hurling himself in desperation against the barricades. But if that were true there was no sign of lack of moral courage and self-respect. There was nothing sullen in his make-up; and if decisions were made contrary to his judgment he was ready to do his best to mitigate the ill-effects he had foreseen.

In the next section the case is made that Frank Barnwell set the gold standard for the intellectual methodology for conceptual airplane design. Barnwell never thought of himself in this regard. But in the light of history, and as a result of this author's research on the history of the evolution of the intellectual design methodology, the case is very strong indeed.

BARNWELL'S DESIGN METHODOLOGY

The history of aeronautical engineering is replete with technological breakthroughs.[10] You can experience some of these breakthroughs simply

[10] Anderson, John D., Jr., *The Airplane: A History of Its Technology*, American Institute of Aeronautics and Astronautics, Reston, Virginia, 2002.

by riding on a modern jet airliner, or as was my good fortune, to once fly the Concorde supersonic transport across the Atlantic Ocean at Mach 2. The advancement of aeronautical technology over the past century is clearly seen by anybody who has lived through this period and is nothing short of phenomenal. What is not so clearly seen, however, is the intellectual thought and mental breakthroughs that drove this advance in technology. The intellectual evolution of the methodology of conceptual airplane design is even more camouflaged by the exponential growth in aeronautical technology. But a close examination of Frank Barnwell's design methodology provides a perspective which is absolutely pivotal to understanding subsequent advances in airplane design. This is the purpose of the present section.

Return to Figure 2.7, showing the essence of the modern intellectual process for conceptual airplane design. The Wright brothers' design methodology discussed in Chapter 2 vaguely resembles the seven intellectual pivot points for conceptual design in Figure 2.7. In fact, to be as successful as they were, they had to follow a somewhat similar thought process. However, they never clearly explained or recorded their precise methodology; one has to extract it piece by piece from their notes and correspondence, and as emphasized in Chapter 2, the pieces do not always neatly fit together. There is no book on airplane design written by the Wrights, although in some sense they "wrote the book" when they achieved their successful flights on December 17, 1903. In some sense it is easy to understand why they did not publish their detailed methodology. When the Wrights realized that they were indeed inventing the first successful airplane, they became very close to the chest with their information, and I am certain they felt that their design methodology was proprietary.

Frank Barnwell had exactly the opposite philosophy. He readily shared his intellectual property with the public. His early success as an airplane designer with the British and Colonial Aeroplane Company gave him confidence in the viability of his brand of methodology for what we today identify as conceptual airplane design. He gave presentations on airplane design to students at British universities, and published numerous papers on the general subject in the leading aeronautical publications such as *Flight*, and *The Airplane*. He was an active member of the Aeronautical Society of Great Britain (soon to become the Royal Aeronautical Society at the end of the war). In their *Aeronautical Journal* in 1919, he published a particularly detailed and lengthy paper entitled "Some Points on Aeroplane Design" full of detailed points on aerodynamics, structures, and stability.[6] Of most importance, however, was his book *Aeroplane Design*[11] published in 1916. (This was published in tandem with a second, but totally separate book, *A Simple Explanation of Inherent Stability* by W. H. Sayers, both books being sandwiched between the same cover.)

The basic contents of *Aeroplane Design* originated as a paper that Barnwell read to the Glasgow University Engineering Society during the winter of 1914. At the encouragement of his friend C. G. Grey, the outspoken editor of the British periodical *The Aeroplane*, they were republished in book form in 1916. In the preface to Barnwell's book, Grey summarized the state of airplane design at that time:[12]

It seems well to make clear why these two writers (Barnwell and Sayers) should be taken seriously by trained and experienced engineers, especially in these days when aeronautical science is in its infancy, and when much harm has been done both to the development of aeroplanes and to the good repute of genuine aeroplane designers by people who pose as 'aeronautical experts' on the strength of being able to turn out strings of incomprehensible calculations resulting from empirical formulae based on debatable figures acquired from inconclusive experiments carried out by persons of doubtful reliability on instruments of problematic accuracy.

Commenting on the impact of Barnwell's book on other airplane designers, Grey goes on to write about Barnwell's methodology:[13]

It is to be noted that his general method of design is approved by other aeroplane designers who have been successful in producing efficient and effective aeroplanes. Consequently the new arrival in the aircraft industry may take it that he is fairly safe in following that method.

Barnwell's methodology parallels six of the seven intellectual pivot points for conceptual airplane design shown in Figure 2.7. He starts with a need for the precise understanding of the requirements for the new design, commercial, military, or other. In his day, these might be as straightforward as, for example, the design of a two-place machine with stipulated minimum and maximum speeds, rate-of-climb, time-to-climb to altitude, range, and payload. He states:[14]

The designer is generally required to produce a machine to carry a certain number of people, petrol and oil for so many hours' flight at full power, a certain weight of observing instruments, perhaps some weapons of offense, fully loaded to be able

[11] Barnwell, F. S., *Aeroplane Design*, McBride, Nast and Co., London, 1916.
[12] Ibid. p 4. [13] Ibid. p.5. [14] Ibid p. 36.

to fly at not less than a certain maximum, and not more than a certain minimum speed, and to climb at not less than a certain minimum rate.

He continues with an example:

Probably the simplest course to take in this brief outline of designing methods is to assume a certain set of conditions has been given and see how we should set about trying to fulfill it. We shall assume, therefore, that we are asked to design a machine to carry two people, pilot and passenger, to fly at 80 m.p.h. maximum and 40 m.p.h. minimum to climb at 7 feet per second fully loaded, to carry petrol and oil for 4 hours, to have a good range of view downwards for the passenger, to carry a full outfit of instruments, i.e., barograph, compass, map case, watches, engines, revolution counter, air speed indicator, inclinometers, etc.

(Recall that Barnwell is writing this in the period around 1914–1915; his list of items is indicative of the primitive state of cockpit instrumentation at the time.)

He then takes the next step, a first estimate of the weight of the airplane. "We are now in a position" he says, "having been given certain requirements, to make a first estimate of weights, deciding in so doing upon the motor to employ."[15] He presents a full page, neatly hand lettered, of a sample calculation for the total weight, including a preliminary choice of an existing engine for propulsion.

With a first estimate for the weight, he moves to step three in Figure 2.7, an estimation of the critical performance parameters. Making a choice of the airfoil shape, he obtains the maximum lift coefficient from wind-tunnel data. From similar data, he estimates the lift-to-drag ratio, and chooses a reasonable wing loading (the ratio of weight to wing planform area). This will help to size the airplane for the next step – a configuration layout. The rather incomplete configuration layout that he shows as an example in his book is given in Figure 3.5. At the bottom of this sketch is the table he used to estimate the weight buildup and the location of the center of gravity based on the configuration layout. From this table, he obtains a better weight estimate, step five of the intellectual pivot points for conceptual design in Figure 2.7. He is then better able to carry out a performance analysis, step six, and to properly size the horizontal and vertical tail as well as choose the dihedral angle for the wing for longitudinal, directional, and lateral stability.

Barnwell wraps up his discussion of the methodology by stating that if the performance of the design "be decently over the requirement we can

[15] Ibid p.36.

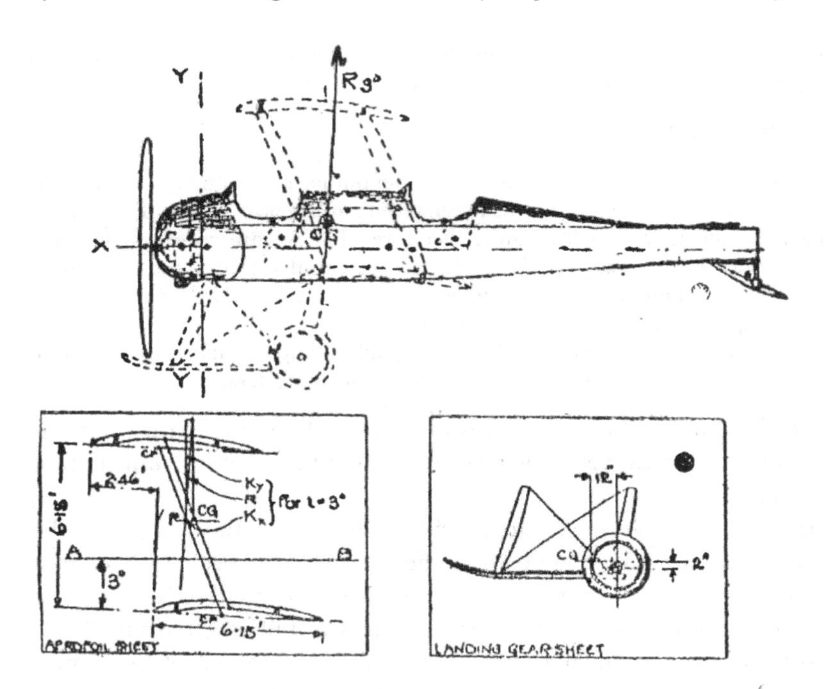

TABLE FOR HORIZONTAL & VERTICAL C.G

ITEM	W	l	h	W×l(+)	W×l(−)	W×h(+)	W×h(−)
Propeller	28	+2·0	–	56	–	–	–
Motor	250	+ ·7	–	175	–	–	–
Cowling	32	+ ·4	+ ·4	13	–	13	–
Motor Mounting	36	– ·2	–	–	7	–	–
Oil & Tank	86	– ·6	+1·3	–	52	112	–
Passenger	175	–2·5	+1·0	–	390	175	–
Passenger's Seat	10	–2·8	+ ·4	–	28	4	–
Petrol & Tank	294	–5·2	+1·4	–	1530	412	–
Body	90	–6·7	–	–	602	–	–
Instruments	30	–7·1	+1·5	–	213	45	–
Controls	30	–7·5	–	–	225	–	–
Pilot	175	–8·7	+1·0	–	1520	175	–
Pilot's Seat	10	–9·1	+ ·4	–	91	4	–
Tail	86	–19·0	+1·0	–	1630	86	–
Tail Skid	7	–19·7	–1·0	–	138	–	7
Aerofoils complete	430	–4·9	+2·8	–	2107	1208	–
Landing Gear	129	–2·5	–3·9	–	322	–	503
TOTAL (loaded)	1898	–4·53	+·91	244	8855	2234	510

W = wgt of Item in lbs
l = Normal dist^ce of CG of Item from line Y–Y. + ahead, – behind
h = " " " " " " " " X–X + above, – below

FIGURE 3.5. A sample configuration layout.
(Barnwell, *Aeroplane Design*, 1916, p. 46.)

FIGURE 3.6. Bristol Bulldog fighter.
(National Air and Space Museum, SI-83-8794.)

consider the preliminary design as finished."[16] This is step six in Figure 2.7, and a "yes" answer to the question in step six is the completion of the conceptual design process as far as Barnwell's world is concerned because the mathematical techniques required for step seven, optimization of the design, were not available at the time.

This is how Frank Barnwell set the gold standard for the methodology of conceptual airplane design in 1916, a methodology that is reflected in the birth of virtually every new airplane to the present day. Barnwell himself designed over 150 different aircraft, many of which, however, remained as paper studies. Setting the gold standard for methodology does not always guarantee a technically successful airplane. During the two world wars, perhaps his best designs were the Bristol Bulldog in the late 1920s (Figure 3.6) and the twin-engine Bristol Blenhem in the late 1930s (Figure 3.7). Due in part to his reserved nature, Barnwell was not a household name, and to the world of aviation he was something of an unsung hero except to a knowledgeable few. But he is one of the most important figures in the history of airplane design. His obituary in *The Aeroplane*[17] stated, "Frank Barnwell was beyond question one of the best airplane designers in this country or in the world. No other designer has turned out so many first-class aeroplanes which have become historic."

[16] Ibid p.71.

[17] Grey, G. C., "Obituary for Frank Barnwell," *The Aeroplane*, 10 August 1938, pp. 162–163.

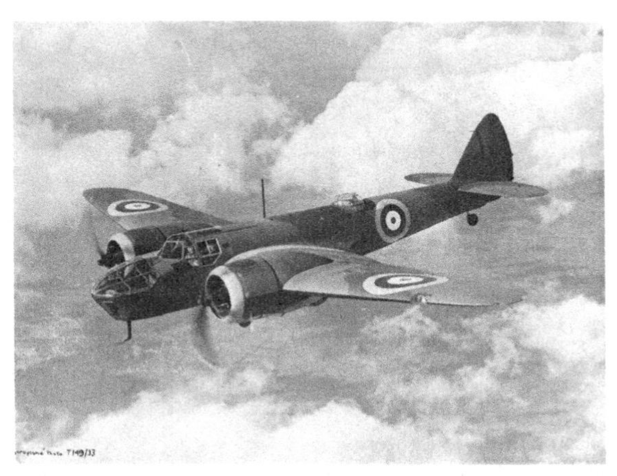

FIGURE 3.7. Bristol Blenheim bomber.
(National Air and Space Museum, NASM-7A07883.)

SETTING A NEW PERSPECTIVE ON THE HISTORY OF AIRPLANE DESIGN

We have come to the crux of the present chapter and perhaps to the book itself. The phenomenal technical advancements that drove the airplane from the era of the strut-and-wire biplane, through the era of the mature propeller-driven airplane, and into the era of the jet-propelled airplane are laid out in Anderson, *The Airplane: A History of Its Technology.*[18] It is easy, therefore, to assume that the very basic intellectual methodology for conceptual design also went through similar advancement, evolving as the airplane evolved. Indeed, this was my earlier preconception. My belief, however, about the intellectual evolution of the methodology of conceptual airplane design has been completely revised by the thinking shared in this chapter. Rather, as argued in this chapter, this basic methodology was set in 1916 by Frank Barnwell, and what we think of as the continued and sometimes spectacular advancements made in airplane design since then have, in reality, been due mainly to the application of *new and advanced technology* hung on the framework of Barnwell's original design philosophy. Airplane advancements during the remainder of the

[18] Anderson, John D., Jr., *The Airplane: A History of Its Technology*, American Institute of Aeronautics and Astronautics, Reston, Virginia, 2002.

twentieth century (streamlining, all-metal construction, the NACA cowl, retractable landing gear, variable-pitch propellers, the jet engine, swept wings, supercritical airfoils, the area rule, etc.) are advancements in technology and do not reflect any fundamental advancement in the intellectual methodology for conceptual airplane design. That methodology was set in 1916 by Frank Barnwell.

I first floated this argument to the aeronautical engineering community in 2006 in the *AIAA Journal*.[19] The response has been quite positive. So the theme taken for the remainder of this book is that the great advances in airplane design, even in Barnwell's time, reflect the use of advanced technology relevant to the time. The remaining "grand designers" discussed in this book were masters of the application of advanced technology, but they did not materially change the basic intellectual methodology of conceptual airplane design as set by Frank Barnwell. Another indicator of the advancement of the technology and its impact on airplane design is that Barnwell's book took 72 pages to cover the technology in 1916, whereas a survey of the airplane technology of today requires more than 700 pages in the leading airplane design text by Dan Raymer.[20]

All this leads us to the next chapter on the "design revolution" of the 1930s, the design of the Douglas DC-3, and the role of its grand designer, Arthur Raymond.

[19] Anderson, John D., Jr., "Airplane Design Methodology: Setting the Gold Standard," *AIAA Journal*, Vol. 44, No. 12, December 2006, pp. 2817–2819.

[20] Raymer, D. P., *Aircraft Design: A Conceptual Approach*, 4th ed., AIAA: Reston, VA, 2006

4

Products of the First Design Revolution

Arthur Raymond and the DC-3

> The pioneers of aircraft design traveled in a wilderness, dependent upon their own resources at every step. No path was marked for them, and the illumination along the way was only such as they themselves might provide. Of research, in its application to aeronautical engineering, there had been hardly even a beginning, and its place had often to be supplied by inspiration, intuition, or empiricism, or go unsupplied.
>
> The student who embarks upon aeronautical studies in 1927 is subject, in some degree, to a contrary embarrassment. Experimental data inundate him. Their ever-growing bulk overwhelms him. The underbrush of obscurity has been cleared away, the field is ever more brightly illuminated by the fruits of governmental, institutional, and private research, yet the very uniformity with which enlightenment is being cast over the subject in its several parts may prove confusing, leaving the exact course to be followed still in doubt.
>
> Edward P. Warner, *Aeroplane Design: Aerodynamics*

The scene: Clover Field, Santa Monica, California, bathed in a gentle breeze blowing in from the ocean, with a comfortable air temperature of 76°F. *The time*: exactly 12:36 pm on July 1, 1933. *The characters*: Carl Cover, vice president of sales and chief test pilot for the Douglas Aircraft Corporation; Fred Herman, a senior designer for Douglas; a crowd of the Douglas employees, including Donald Douglas and his assistant chief designer, Arthur Raymond; and a sleek, aesthetically beautiful new airplane, the first commercial airplane ever designed at Douglas, aptly labeled the DC-1. *The action*: With Cover at the controls and Herman as copilot, the twin-engine DC-1, powered by two Wright Cyclone engines of 710 horsepower each, throttles wide open, roars down the runway, and lifts off for its maiden flight. Douglas and his small company have a great deal riding on the success of this airplane. Designed to meet stringent specifications set down by the Transcontinental and Western Air, Inc. (TWA), the DC-1 embodies the absolute synergistic best of modern aeronautical engineering technology in 1933. If successful, the DC-1 would be, by far, the most economic, most comfortable, highest speed, highest flying, and safest airplane to exist at that time. Jack Frye, vice president of TWA, who had drafted the specifications, is looking to the airplane to revolutionize commercial air travel. Little does anybody know that potential disaster looms on the horizon. As Cover puts the DC-1 in a climb about thirty seconds after takeoff, the left engine quits; a moment later the right engine sputters to a stop. As the airplane noses over, however, the engines start again. Cover starts to climb again, but once again the engines stop. They start again when the nose tips down. For the next ten minutes, Cover puts on a display of expert piloting, coaxing the DC-1 up to 1,500 feet, following a sawtooth flight path alternating between a climb, the engines cutting off, a nose over, the engines starting again, and another climb until the engines quit again. At 1,500 feet, Cover judges that the DC-1 is at a safe enough altitude to allow him to bank and return safely to the runway. A few minutes later, the airplane and its occupants are on the ground unscathed.

But what is wrong? The airplane and its engines appear to be mechanically sound. Five days go by as the engines are taken apart and reassembled more than a dozen times. On the test block, the engines run perfectly. During the process, Carl Cover suggests to the technicians from the Wright Aeronautical Corporation that they look at the carburetors but they object, thinking the problem is elsewhere. On the fifth day, Cover's suggestion is finally carried out. To the amazement of the technicians, the carburetors, which meter fuel to the engine, are found to be installed backwards. In this position, the carburetor floats would cut off the fuel flow when the airplane is in a nose up altitude. The carburetors are quickly rotated 180 degrees. The problem is fixed. On July 7, the DC-1 takes off again, and the engines perform perfectly.[1]

Such is the saga of the beginning of the most successful series of commercial airliners for the next thirty years. Only one DC-1 was built; a photograph of it is shown in Figure 4.1. The DC-1 was essentially an experimental airplane. It was quickly followed by the look-alike but

[1] Anderson, John D., Jr., *The Airplane: A History of Its Technology*, American Institute of Aeronautics and Astronautics, Reston, Virginia, 2002, pp. 183–185.

FIGURE 4.1. Douglas DC-1.
(National Air and Space Museum, NASM-1A38742.)

slightly larger DC-2, of which Douglas was to manufacture 156 in twenty different models that were used by airlines around the world. The DC-2, in turn, quickly evolved into the look-alike but still larger DC-3, which was to become one of the most successful airplanes in the history of flight and aeronautical engineering. When the DC-3 production line was finally shut down at the end of World War II, 10,926 had been built. The Douglas DC-3 was the first widely successful example of the mature propeller-driven airplane. In this chapter we will take a closer look at the airplane, its design, and its grand designer, Arthur Raymond.

THE DESIGN REVOLUTION

Take a closer look at the airplane shown in Figure 4.1. Gone is the strut-and-wire biplane configuration. Gone is the vegetable airplane made from wood and fabric. Gone are the engine cylinders exposed to the airstream, and the fixed landing gear permanently hanging underneath. Gone is the boxy shape of the fuselage. Instead, what we see in Figure 4.1 is a beautiful streamlined all-metal airplane – a monoplane with a high aspect ratio wing, and engines sheathed inside a high technology low drag cowl. The DC-1was a product of the "design revolution" of the late 1920s and well into the 1930s. The design revolution was *not* a revolution in the methodology of conceptual airplane design. Rather, it was a confluence of major *technical* advances in aerodynamics, structures, and propulsion during this period, and the willingness of some airplane designers to incorporate this new technology in the design of new airplanes.

Features of the design revolution are described in great detail by Anderson.[2] The explosive growth of aeronautical research alluded to by E. P. Warner in the quote that begins this chapter made the design revolution possible. (We will have more to say about Ed Warner and his seminal airplane design text later in this chapter.) It is useful for our remaining discussions in this book to summarize briefly some of the main features of the design revolution.

(1) *Streamlining*. Aerodynamicists learned that smooth elongated bodies made with gentle curves resulted in low aerodynamic drag. This was in comparison to the earlier high-drag box-like configurations with flat surfaces and sharp corners. Aerodynamicists learned that the airflow separated from these sharp corners, creating large pressure drag (identified as pressure drag due to flow separation.) When the body is streamlined, the flow remains attached to the body almost all the way to the trailing edge, thus virtually eliminating pressure drag. This leaves only skin-friction drag as the main source of drag. Streamlining as a means to reduce drag is one of the most visible features of the design revolution. Needless to say, it is also an aesthetically beautiful feature, and streamlining was adapted by many Art Deco artists in the design of furniture, household appliances, buildings, etc. The well-respected survey of art history, *Gardner's Art Through the Ages*, elaborates on the connection between streamlining and the Art Deco movement in the 1920s and 1930s.[3] It notes that "Art Deco had universal application – to buildings, interiors, furniture, utensils, jewelry, fashions, illustrations, and commercial products of every sort. Art Deco products have a 'streamlined' elongated symmetrical aspect." It goes on to state that "these simple forms are inherently aerodynamic, making them technologically efficient (because of their reduced resistance as they move through air or water) as well as aesthetically pleasing. Designers adopted streamlined designs for trains and cars, and the popular appeal of these designs led to their use in an array of objects, from machines to consumer products." Finally, it emphasizes that "this streamlined look was integral to Art Deco."

[2] Ibid, p. 183-282.

[3] Kleiner, Fred S. and Mamiya, Christine J., *Gardner's Art Through the Ages*, 12th Ed., Thomas Wadsworth, Belmont, CA, 2005, p. 1014.

A recent study by Dominick Pisano, Curator at the Smithsonian's National Air and Space Museum, looks in detail at the connection between the aerodynamic concept of streamlining and industrial design, making the case that streamlining was directly appropriated from aerodynamics to sell or promote products in the midst of the Depression.[4]

The airplane design community had an abrupt awakening to the aerodynamic virtue of streamlining by the famous British aeronautical engineer and Cambridge University professor, Sir B. Melvill Jones. Jones made a resounding call for streamlining in a watershed paper entitled "The Streamline Airplane" delivered to the Royal Aeronautical Society in 1929.[5] Jones noted a characteristic typical of airplanes in the 1920s, namely, that the power required by an airplane to overcome head resistance was 75 to 95 percent of the total power used. Head resistance, a term deriving as far back as the 1890s and Octave Chanute's *Progress in Flying Machines*,[6] is the sum of skin-friction drag and pressure drag due to flow separation (frequently called form drag). Because little could be done to reduce skin-friction drag, except to reduce the exposed surface area of the airplane, the primary target reduction had to be form drag. "We all realize," stated Jones, "the way to reduce this item in the power account is to attend very carefully to *streamlining*." Jones went on to define what he called the "perfectly streamlined airplane" as one where the form drag was reduced to zero.

The aspect of Jones's paper that most shocked airplane designers into greater awareness was his plot of horsepower required versus velocity, which compared Jones's ideal "perfectly streamlined" airplane with various real airplanes of that time. This plot is so important to our story of airplane design that an exact facsimile is given in Figure 4.2. The solid curves at the bottom of the graph show the power required for the ideal airplane. (They take into account only skin-friction drag and the induced – vortex – drag due to the vortices streaming downstream of the wing tips.) The different solid curves correspond to different wing spans and weights. The solid symbols are data points for real airplanes. Jones pointed out that the vertical

[4] Pisano, Dominick A., "The Airplane and the Streamline Idiom in the United States," paper presented at the meeting of the Society of Architectural Historians, Denver, CO., 23–27, April, 2003.

[5] Jones, B. Melvill, "The Streamline Airplane," *Aeronautical Journal*, Vol. 33, 1929, pp. 358–385.

[6] Chanute, Octave, *Progress in Flying Machines*, 1894. Reprinted by Lorenz and Herweg, Long Beach, CA, 1976.

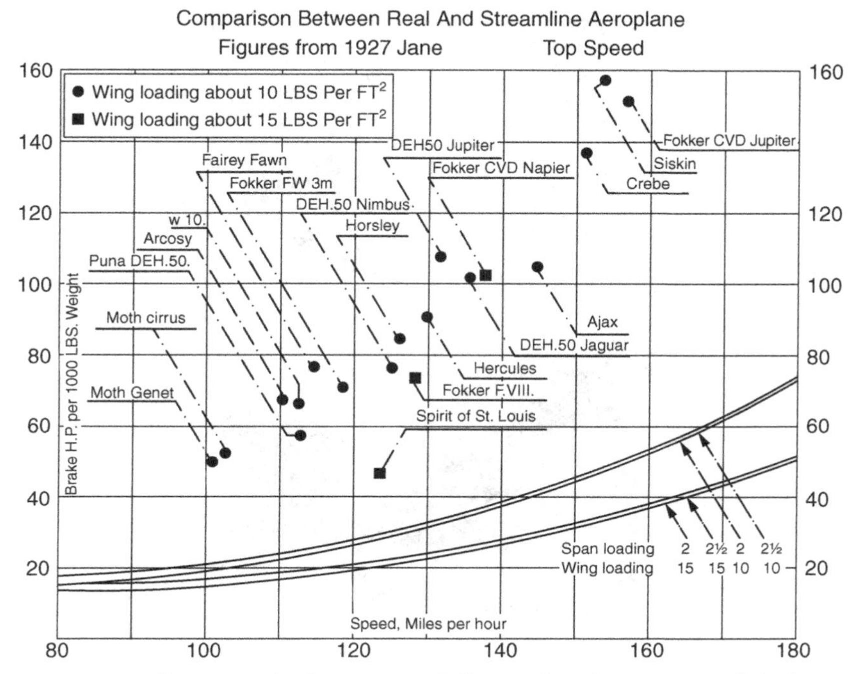

FIGURE 4.2. Power required versus speed. Comparison between actual airplanes circa 1927 and Melvill Jones' ideal streamlined airplane.
(From "The Streamline Airplane," *Aeronautical Journal*, 33 (221), 1929.)

distance between any of these data points and the solid "ideal" curve was the power expended by the real airplane in overcoming form drag, and that "unnecessary" power consumption was considerable for all of the airplanes listed. Another way of interpreting Jones' graph is to examine the horizontal distance between a given data point for a real airplane and the ideal curve. This represents the *increase in velocity* that could be achieved at a given power available if there were no form drag. For example, choosing the data point for the Argosy, a typical trimotor biplane transport in the 1920s, we see that for no form drag, the top speed of the Argosy would have been a blistering 175 mph rather than the actual value of 110 mph.

Jones' paper knocked the socks off the airplane design community. After that, streamlining became a major part of the design revolution in the 1930s. A graphic example of the effect of streamlining is seen in Figure 4.3, showing the S.E.5 of World War I morphing into the beautifully streamlined Spitfire of World War II.

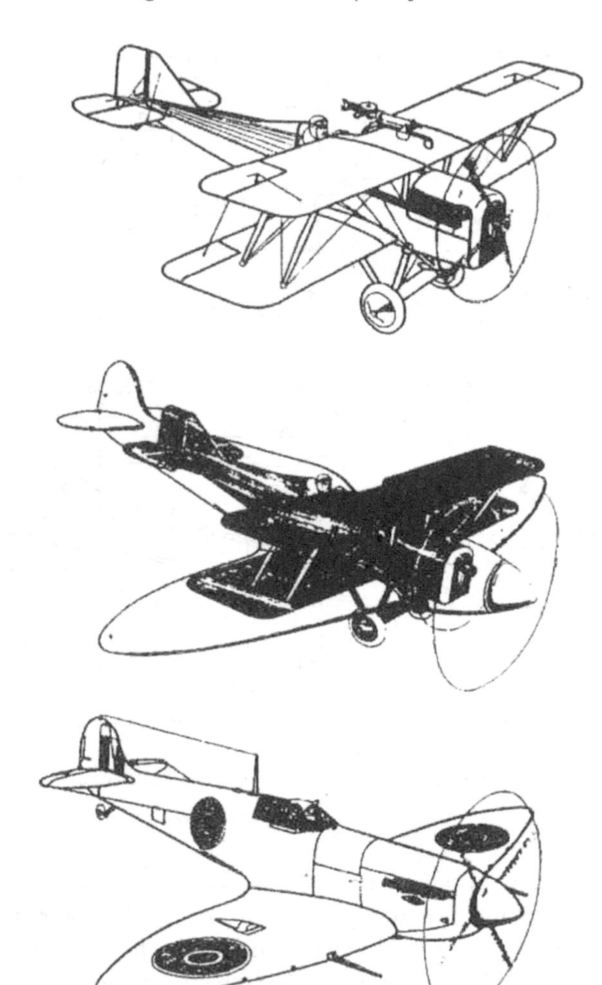

FIGURE 4.3. Morphing of the World War I SE5 into the World War II Spitfire. (From a presentation to the Institute of Aeronautical Sciences by William S. Farren, 1944.)

(2) *The NACA Cowling.* In 1926, airplanes could be divided into two general categories on the basis of the types of engine used – the liquid-cooled in-line engine or the air-cooled radial engine. The former was generally enclosed within the fuselage and did not present much of a problem in regard to streamlining. The latter relied on the airflow over the cylinders to cool the engine and thus, the cylinders, arrayed like

spokes on a wheel, were directly exposed to the airstream. As a consequence, air-cooled radial engines created a lot of drag. However, radial engines had several advantages that led to their use in many airplane designs: lower weight per horsepower, fewer moving parts, and lower maintenance costs. The US Navy was partial to air-cooled radial engines because they continued to perform well despite the jarring impacts of carrier landings. Therefore, it is no surprise that in June 1926, the chief of the US Navy's Bureau of Aeronautics requested that the NACA study how a cowling could be wrapped around the cylinders of radial engines, designed so as to reduce drag without interfering with cooling capacity. This set into motion one of the important aerodynamic developments in the era of the mature propeller-driven airplane – the NACA cowling.

The idea of cowlings was not new. For example, the Bristol M.1C (Figure 3.4) had a cowling wrapped around a LeRhone rotary engine. Many of the rotary engines used on World War I airplanes were housed inside curved metal cowlings. Those cowling designs were mainly intended to collect the oil thrown out of the rotating engine, and their shapes were based more on art than on science or any sound knowledge of aerodynamics. Fortunately, cooling was not a problem for those cowled rotary engines because the cylinders were always rotating through the air inside the cowling. The problem arose with the stationary radial engines that became prevalent in the 1920s. For these, the only way to have air cooling the cylinders was to expose the cylinders to the main airstream in flight.

The NACA cowling research was the first major test program to be carried out in the newly operational Propeller Research Tunnel (PRT) at the NACA Langley Memorial Laboratory. This large tunnel had a test section 20 feet in diameter, large enough in which to mount a full-size fuselage with an installed engine and propeller, and a maximum airspeed of 110 mph. Fred Weick, a relatively young aeronautical engineer from the University of Illinois, had just become director of the PRT. Weick was given responsibility for the NACA cowling program because the PRT was the logical place to carry out the research.

For the next ten years the NACA carried out research on cowlings, most of it experimental. Within a year after beginning the program, Weick and his associates had demonstrated that a properly designed cowling could dramatically reduce the form drag associated with radial engines without adversely affecting engine cooling – findings that were immediately snapped up by industry and incorporated into new airplane

FIGURE 4.4. Lockheed Vega.
(National Air and Space Museum, SI-2009-7961.)

designs. The cowlings seen on the engines of the DC-1 in Figure 4.1 are an example of the very successful NACA design. For some airplanes, the NACA cowling reduced the drag by as much as 60 percent compared with the case of totally exposed cylinders. Indeed, in the wind-tunnel tests, the cowling reduced the drag almost back to the value measured without the engine installed. The famous Lockheed Vega was first produced in 1927 with its cylinders exposed to the airstream, as was conventional at the time. Its maximum speed was 165 mph. In 1929, with the NACA cowling added, its maximum speed was dramatically increased to 190 mph. The Vega (Figure 4.4) became one of the first commercially built airplanes to use the NACA cowling. It was flown by pilots such as Wiley Post and Amelia Earhart.

In 1929 the NACA cowling won the Collier Trophy, an annual award commemorating the most important achievement in American aviation. The NACA cowling became an essential element of the design revolution.

(3) *Reliable, More Powerful Engines.* Liquid cooling, or air cooling? That was the question facing the airplane engine designer after World War I. The answer was both. The era of the mature propeller-driven airplane saw the development of powerful liquid-cooled engines, most of which were used in military airplanes – especially fighter airplanes where streamlining was all important. The same era saw the development of highly successful air-cooled radial engines, which by the end of World War II were by far the dominant power plants for most airplanes. These engines, in combination with the development of the viable-pitch propeller, constitute a vital part of the design revolution.

Hispano-Suiza pioneered the development of liquid-cooled engines for airplanes in France during World War I. The famous SPAD VII and SPAD XIII fighters were powered by Hispano-Suiza liquid-cooled engines producing up to 220 horsepower. After the War, Curtiss in the United States and Rolls-Royce in Britain took up the cause. Curtiss produced the D-12 engine that, with its 325 horsepower, powered a series of Curtiss racing airplanes to victory in the 1920s. Rolls-Royce took the basic Hispano-Suiza design and evolved it into a series of designs that led to the famous Merlin engine of the 1930s. The Supermarine S-6B was powered to victory in the final Schneider Cup race on September 13, 1931 by a point-designed Rolls-Royce R liquid-cooled engine that produced 2,300 horsepower for limited durations; the S-6B averaged a winning speed of 340.1 mph. Later that month, the world's speed record of 401.5 mph was set by the same S-6B.

The development of air-cooled radial engines, unlike that of the liquid-cooled engine, was not a seamless, continuous series of developments from World War I radials. The major type of air-cooled engine from that period was the rotary engine. The many disadvantages of having the whole engine rotate about a fixed crankshaft made the rotary unpopular by the end of the war. In 1918, the liquid-cooled engine reigned supreme, and the air-cooled engine appeared to become an artifact of history.

Not for long, however. On the scene came two men, Samuel D. Heron, originally in England, and Charles L. Lawrance in the United States. Heron, an apprenticed mechanic, had worked for Rolls-Royce, Napier, and Siddeley aircraft engine companies before being hired by the Royal Aircraft Factory at Farnborough, England, during World War I. There he was involved in the design of the first successful aluminum air-cooled cylinders. This work eventually led to the British Jupiter air-cooled radial engine in 1922 which produced about 400 horsepower. Large numbers of the Jupiter were manufactured in England and Europe and were employed mainly on military airplanes. The Jupiter had a problem, however. Because of poor exhaust valve cooling, the engine required frequent maintenance. Heron once said about the Jupiter that its consumption should be given in terms of pounds of exhaust valves, rather than in pounds of fuel, per horsepower-hour![7] In 1921, Heron came to the United States, where he worked for the US Army Aeronautical Research and Development Laboratory at McCook Field in Dayton during the early

[7] Anderson, John D., Jr., *The Airplane: A History of Its Technology*, American Institute of Aeronautics and Astronautics, Reston, VA, 2002, p. 248.

1920s. It was at McCook Field that Heron pioneered the liquid-filled valve concept for cooling: he had some success using a mixture of sodium and potassium nitrate. By 1928, Heron's sodium-filled valve had been adopted on high performance engines. It was a major contribution to engine technology and was one of the primary reasons for the improvement of engine reliability and endurance – another hallmark of the design revolution.

At the same time that Heron was working on air-cooled cylinders in England, Charles L. Lawrance was working to develop a practical air-cooled engine in the United States. Lawrance was a New Englander, with a B.A. degree from Yale in 1905 and a *Diplôme École des Beaux Arts*, Paris, in architecture in 1908. While he was in Paris, he became interested in aircraft engines. On returning to the United States at the start of World War I, Lawrance formed a small company, the Shinnecock Airplane Company, for the purpose of producing a light private airplane powered by a small air-cooled engine that he designed. In 1917 he founded the Lawrance Aero-Engine Corporation and began to design air-cooled engines for the US Army and the US Navy. In 1921 Lawrance created the popular and successful 9 cylinder, 200 horsepower J-1 air-cooled radial engine. The Navy ordered 200 of the engines, starting a long duration love affair between the Navy and air-cooled radial engines. As already mentioned, these engines had fewer moving parts and were less susceptible to impact damage during carrier landings than liquid-cooled engines. The Navy became the champion of air-cooled radials. Unfortunately, Lawrance's company was small and did not have the mass production facilities desired by both the Army and the Navy. With the Navy's explicit encouragement, Lawrance's company was bought by the Wright Aeronautical Corporation. Gradually, and initially reluctantly, Wright began to shift away from its earlier line of liquid-cooled engines to focus on air-cooled radials. The Wright Aeronautical Corporation built improved versions of Lawrance's engine, known in the 1920s as the Wright J-3 and J-4.

Sam Heron left McCook Field in 1926 to join the Wright Aeronautical Corporation, of which Lawrance was now president, hence bringing together the two pioneer developers of the air-cooled radial engine. The immediate joint product of these two men was the Wright J-5 Whirlwind, one of the most successful engines in the 200 horsepower class. The Whirlwind was used by Charles Lindbergh in his Spirit of St. Louis and it won the coveted Collier Trophy in 1927 for aviation's most important development of the year. With experience from the Whirlwind, Lawrance designed a completely new engine, the Wright Cyclone, which in its various forms grew from 500 horsepower in the early 1930s to 1,900 horsepower by the beginning of World War II. Wright Cyclone engines powered most versions of the Douglas DC-3.

Pleased with the success of air-cooled radial engines in the 1920s, the Navy felt that it needed a second source of supply. With the promise of engine orders from the Navy if they produced a suitable engine, a small group of Wright Aeronautical employees – led by the then president of the company, F. B. Rentschler – left to form the Pratt and Whitney Aircraft Company in Hartford, Connecticut. By 1926, the fledging Pratt and Whitney group produced the first of a highly successful line of Wasp air-cooled radial engines. During the design revolution of the 1930s, Wright and Pratt and Whitney would be the two primary suppliers of powerful and reliable air-cooled radial engines.

(4) *Variable-Pitch Propeller, Supercharging, and 100-Octane Fuel.* Three other propulsion-related technical advancements came together during the design revolution. The first is the development of the variable-pitch propeller. A propeller is the device that converts the power from a reciprocating engine to forward thrust, the actual force that propels the airplane forward. The propeller is a purely aerodynamic device. It is essentially a twisted wing. The twist is necessary because each section of the rotating propeller is moving at a different speed; the blade sections near the tip are moving faster than those sections near the hub. (If you are sitting on a rotating carousel, your speed near the outer rim is much faster than if you are sitting near the hub at the center.) At a given section of the propeller, the direction of the local airflow is the combination of the forward motion of the airplane and the local velocity due to the pure rotation of the propeller. Hence, if the propeller was not twisted (a flat blade for example) the local angle of attack to the *local* airflow seen by each section of the blade would be different. Since the maximum lift-to-drag ratio (maximum aerodynamic efficiency) of an airfoil occurs at a specific optimum angle of attack, the only way for each section of the blade to see this optimum angle of attack is to properly twist the propeller blade. Furthermore, the angle made by a given airfoil section relative to the plane of rotation (the propeller "disk") is called the pitch angle; the pitch angle of each section of a given twisted propeller changes from the hub to the tip. We can average all the different pitch angles for all the different airfoil sections and define an overall pitch of the propeller.

The propellers used by airplanes prior to the 1930s were rigidly fixed to the engine shaft, i.e., they were *fixed-pitch propellers*. This was not a problem at the low flight speeds and altitudes for World War I era airplanes. But for more modern airplanes with a greatly expanded range of speed and altitude, to maintain optimum propeller efficiency the pitch should be changed during the flight. To design a mechanical mechanism to do this is easier said than done. The value of a *variable-pitch propeller* was recognized even before World War I. The War provided a mild stimulus for work on variable-pitch propellers, but due to massive mechanical complexities, no practical solutions came from these efforts. The successful solution to the problem did not happen until the metal propeller was developed, and both hydraulic and electric mechanisms were successfully designed for changing the pitch. These developments finally came together in the 1930s, just in time for the higher performance airplanes that were products of the design revolution. One person stands out as the principal developer of the first successful variable-pitch propeller, Frank W. Caldwell. The story of his persistence, technical knowledge, and innovative ideas is an inherent part of the history of the design revolution. The story of Caldwell's life and his contributions to aeronautical technology is nicely told by Kinney.[8,9]

The development of a practical supercharger was another propulsion-related contribution to the design revolution, although it came rather late in the period. The power output of a reciprocating engine varies directly as the density of the ambient air surrounding it. For example, an aircraft engine that produces 1,000 horsepower at sea level will produce only about 500 horsepower at an altitude of 22,000 feet, where the air density is only half that at sea level. This basic fact had been recognized since the time of the Wright brothers, and for the first twenty years of powered flight, the progressive power loss with increasing altitude was simply accepted. But with the other technical advancements being made during the design revolution, this situation could no longer be tolerated. Indeed, something can be done to maintain engine power with altitude – supercharging. A supercharger is a mechanical device that compresses the air before entering the engine manifold; it is usually in the form of a centrifugal blower that is driven from the main engine via a gear connection to the

[8] Kinney, J. R., "Frank W. Caldwell and Variable-Pitch Propeller Development," *Journal of Aircraft*, Vol. 38, Sept.-Oct. 2001, pp. 967–976.

[9] Kinney, J. R., *Reinventing the Propeller*, Cambridge University Press, 2017.

main crankshaft (a gear-driven supercharger), or by a turbine inserted in the exhaust gas stream from the engine (a turbocharger). The practical supercharger took a long time to develop. For a brief history, see Anderson.[10] It's sufficient to say that geared superchargers came into use in 1927, and after 1930 almost all military and civil transport airplanes were powered with engines equipped with geared superchargers. Practical turbochargers took longer to enter service. The primary stumbling block was finding suitable materials for the turbine wheels that had to endure the high-temperature exhaust gases. In March 1939, however, turbochargers were successfully tested on a Boeing B-17 bomber and finally accepted by airplane designers.

Finally, one of the lesser known but nevertheless important developments in propulsion technology during the 1930s was the advent of high-octane fuels. These fuels allowed engines of a given size to produce more power and were one of the major contributions to the rapid increase of engine power during the period of the design revolution.

The power output of a reciprocating engine is dependent on the pressure ratio achieved during the compression stroke because the higher the pressure the more efficient the combustion of the fuel–air mixture. If the pressure is too high, however, the combustion process, instead of being a well-behaved controlled burning mechanism, will instead be a detonation that is less efficient and that can damage the pistons and cylinders – an audible phenomenon called "pinging" or "knocking." The problem of knocking limits the allowable design compression ratio of the engine, hence limiting the design power output of the engine.

In the 1920s, two developments occurred that improved the situation. In 1921, Thomas Midgley, Jr., of General Motors discovered that tetraethyl-lead used as an additive to gasoline improved the antiknock properties of the fuel. It was adopted as an additive in 1926 by the US Army and in 1927 by the US Navy. Of more importance, however, was the discovery of the effect of isooctane on knocking by Graham Edgar of the Ethyl Corporation in 1926. Isooctane, a paraffin with the chemical formula of C_8H_{18}, is a normal product of the distillation of petroleum. The octane rating of gasoline is the amount of isooctane present by volume. In the 1920s, the octane ratings of aviation fuel were essentially that for automobiles at the time – about 50-octane, or 50 percent isooctane by volume. Edgar discovered that with more isooctane present in the fuel, the

[10] Anderson, pp. 250–253.

compression ratio could be made higher before knocking occurred. The Army quickly adopted 87-octane gasoline in 1930, and by 1937 it was the standard for civil aviation as well. Engineers at the Army's Wright Field in Dayton, however, pushed for more. As a result, the Army adopted 100-octane fuel as the military standard in 1936. This fuel became the norm for the military during World War II; it was one of the factors that gave Allied airplanes during World War II a technical advantage because Germany and Japan continued to use the lower octane aviation fuel.

(5) *Structures and Materials.* The all-metal airplane came into its own at the beginning of the design revolution. Hugo Junkers in Germany designed and built the first practical all-metal airplane in 1915, the Junkers J.1. Although other all-metal airplanes made from a special alloy called Duralumin were designed in the decade following Junkers' pioneering J.1, this type was not generally accepted by the airplane design community at the time. Three advances, one in materials and two in aircraft structures, changed this situation, all occurring during the 1920s.

The advance in materials involved the protection of aluminum alloys from corrosion. The concern over corrosion was a major impediment to the adoption of Duralumin in both Britain and the United States, in spite of the fact that Junkers forged ahead with his successful all-metal airplanes of World War I. This objection to aluminum was gradually removed when Guy Dunstan Bengough and H. Sutton, working for the National Physical Laboratory in the United Kingdom in the mid-1920s, developed a technique of anodizing aluminum alloys with a protective oxide coating, and when E. H. Dix in the United States in 1927 discovered a method of bonding pure, corrosion-resistant aluminum to the external surface of Duralumin. The latter, in manufactured form, is called Alclad. The NACA, after carrying out corrosion tests on Alclad, gave its approval, and Alclad was accepted by the US Army and the US Navy for new all-metal aircraft. The first major civil airplane to use corrosion resistant aluminum was the pioneering Boeing 247, which was fabricated from Alclad 17.ST.

The two major advances in structural design took place in Germany. The first was the practical introduction of the stressed-skin concept. Junkers had designed his all-metal airplanes with the idea that the metal skin would carry part of the load. Except for a few exceptions, however, Junkers used corrugated metal to obtain stiffness of the skin. The corrugations added extra surface area, thus increasing the skin-friction drag on the airplane; moreover, the corrugations interfered with the smooth aerodynamic flow over the surface. The practical use of a smooth stressed-skin structure is largely due to Adolf Rohrback. Born in 1889 in Gotha, Germany, Rohrback received his engineering diploma from the Technische Hochschule Darmstadt and went on to earn his doctorate in engineering from the Technische Hochschule Berlin-Charlottenberg in 1921. Afterward, he established the Rohrback Metal Airplane Company in Copenhagen, Denmark, the location chosen to circumvent the strict restrictions imposed on the German aircraft industry by the Treaty of Versailles. He concentrated on the design and construction of all-metal flying boats. In contrast to Junkers' airplanes of corrugated aluminum, Rohrback's airplanes had smooth metal skins, hence lower overall friction drag. Moreover, the internal structure of Rohrback's wings involved a strong, metal box beam with a rounded nose in front and a tapered section in back, all made from aluminum. This design is shown schematically in Figure 4.5, which contrasts the early evolution of metal wing structures. Rohrback's wing design was considered revolutionary at the time; later it served as a model for the stressed-skin wing structures that have dominated airplane design from 1930 to the present.

Rohrback's rectangular skin panels were fastened to a frame which supported the panels on all four sides. Prevailing practice at the time dictated that the frame plus panel combination should be strong enough to prevent the panel from buckling because buckling was viewed as structural failure. This laid the groundwork for the second major advance in metal aircraft structures, namely the discovery that a structure of mutually perpendicular members covered with a thin skin did not fail if the skin buckled. Herbert Wagner, an engineer working for Rohrback, made this discovery in 1925, although he did not publish his findings until 1929. When the skin is allowed to carry the maximum possible load, the spacing between frames could be increased, resulting in lighter-weight structures.

Without knowledge of Rohrback's or Wagner's work, Jack Northop in the United States pioneered a multicellular wing structure of spars, ribs, and stringers covered with smooth, stressed-skin sheets. He independently discovered that the sheets did not fail even after they began to buckle. The Northrop wing was adopted for several aircraft, including the DC-1–2-3 series. The DC-3 wing structure is shown in Figure 4.5. This wing structure proved to be exceptionally long lasting. Each of the many load

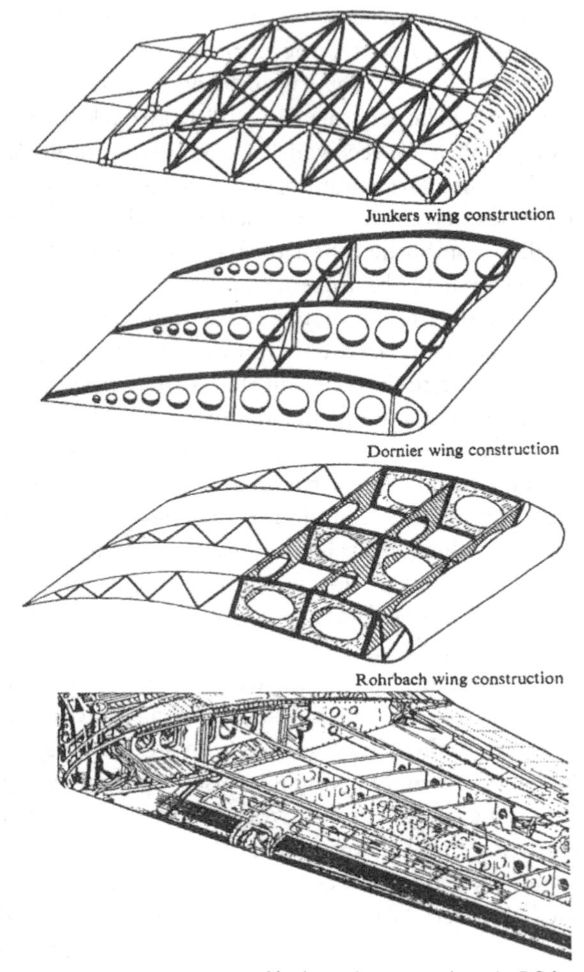

FIGURE 4.5. Four different structural designs for wings.
(From R. Miller and D. Sawyers, *The Technical Development of Modern Aviation*, Praeger Publisher, New York, 1970.)

carrying elements of the wing carried lower stress, which greatly increased the fatigue life of the DC-3.

In short, the all-metal, mostly aluminum, stressed-skin aircraft structure became the norm during the design revolution. The all-vegetable airplane became a thing of the past.

THE DESIGN REVOLUTION: THE BOTTOM LINE

The design revolution was a revolution in *technology* applied to airplanes. There was *no* revolution in the intellectual *methodology* of conceptual airplane design. All the advances in technology discussed in the previous section were just that – advances in technology. The products of the design revolution – the airplanes with marked improvements in performance, reliability, and safety – were a result of this advanced technology hung on the skeleton of the intellectual methodology for conceptual airplane design developed by Frank Barnwell in 1916, as discussed at length in Chapter 3.

In this and the next chapter, we look at some of the airplanes of the design revolution, and at their designers. We continue to explore the ways of thinking and the special attributes that made these designers great.

NOT AN AIRPLANE, BUT A BOOK: THE INFLUENCE OF EDWARD P. WARNER ON AIRPLANE DESIGN

The design revolution was fueled in part by a cadre of young aeronautical engineering graduates from a growing number of universities offering degree programs in the field. By 1936, at least a dozen major institutions of higher learning in the United States offered bachelor's and advanced degrees in aeronautical engineering. Six of these had benefitted greatly from grants from the Daniel Guggenheim Fund for the Promotion of Aeronautics.[11] Set up in 1926, this fund made substantial grants to establish or enhance schools of aeronautical engineering at six universities: Stanford, The California Institute of Technology, University of Washington, University of Michigan, MIT, and the Georgia School of Technology (now the Georgia Institute of Technology). A year earlier, Daniel Guggenheim had made a personal gift of $500 thousand to New York University to establish the Daniel Guggenheim School of Aeronautics.

Historically, the first college level courses in universities were taught at MIT beginning in 1913; they were graduate level courses taught by the Assistant Naval Constructor Jerome C. Hunsaker through the auspices of MIT's Naval Architecture Department. In 1916 the University of Michigan became the first university in the United States to establish an

[11] Hallion, Richard P., *Legacy of Flight: The Guggenheim Contribution to American Aviation*, University of Washington Press, Seattle, 1977.

undergraduate curriculum leading directly to a bachelor's degree in aeronautical engineering. Thus, before 1920, students were graduating with college degrees in aeronautical engineering.

One of the early students in MIT's aeronautics program was Edward P. Warner (1894–1958), who had come to MIT in 1916 fresh from earning a B.A. with honors from Harvard. A year later he obtained his B.S. degree from MIT, followed by an M.A. degree in 1919. Warner was exceptionally gifted intellectually, so much so that MIT put him in the classroom as an instructor of aeronautics during the first year of his graduate studies. After graduation, Warner joined the fledgling NACA's Langley Memorial Laboratory in Hampton, Virginia, where he was made chief physicist and served as head of aerodynamic and flight research. He designed the first wind tunnel at Langley and oversaw its construction. In 1920 he returned to MIT as an associate professor of aeronautical engineering and he taught there for the next six years.

During this period, Ed Warner wrote a seminal textbook on airplane design. Simply titled *Airplane Design: Aerodynamics*, it was published by McGraw-Hill in 1927. I have read this book from cover to cover, and more than eighty years after its publication it still stands as one of the clearest and most logical presentations I have ever read on the subject. In 1927 it provided new college students as well as practicing professionals with the most valuable reference source and teaching instrument on airplane design at that time. This book touched and motivated many students who would later participate in the design revolution. Much later one of Warner's students at MIT, T. P. Wright, who later became the vice president for engineering at the Curtiss-Wright Aeronautical Corporation, said about the book:[12] "This rather classic text served as a 'bible' in this field to many of us. Many observations in it were ahead of the times. In subsequent years, what appeared to me as a new concept, would upon search, be found in *aerodynamics* worked out in detail but not fully appreciated nor absorbed on original reading." Warner had an encyclopedic mind, amazing his friends and acquaintances with his prodigious knowledge. He was technically brilliant, and he could concentrate on the matter at hand to the exclusion of all others. "I'm afraid we didn't absorb too much from his rapid-fire lectures," Wright noted, "but we were tremendously impressed by his knowledge and fluency." Warner's secretary at MIT, Miss Elizabeth Brown, said

[12] Wright, T. P., "Edward Pearson Warner: An Appreciation," *Journal of the Royal Aeronautical Society*, Vol. 62, October 1958, pp. 31–43.

about him: "We used to call him a 'dynamo' in trousers at Tech. He always went around with his pockets bulging with papers. He had special pockets made in his suits for this purpose."[13] Another of his students, Paul Johnston, who later became the Director of the Institute of Aeronautical Sciences, reflected about Warner:[14] "He usually had an armful of papers and magazines wherever he went and he always had a line of multi-colored pens, pencils, slide rules, and so on, in his pockets." Such was the man who provided the aeronautical engineering profession with the 1927 textbook that clearly set the tone for the design revolution in the 1930s.

We observe, however, that Warner's book says nothing about the intellectual methodology of conceptual airplane design. Rather, in keeping with the nature of the design revolution, *Airplane Design: Aerodynamics* is a clear and voluminous presentation of aeronautical *technology* – the type of technology that was easily hung on the skeleton of the intellectual methodology as established by Frank Barnwell in 1916.

Edward Warner left MIT in 1926 to become the Assistant Secretary of Navy for Aeronautics, thus beginning a long and distinguished career in public service and expanding his influence to spheres of aeronautics policy, economics, safety, and international cooperation, among others. Holding numerous prestigious positions in government and regulatory agencies, he served the last twelve years of his professional life as president of the International Civil Aviation Organization (ICAO), the organization that was instrumental for establishing uniform international rules and procedures on communications, traffic control, licensing, aerology, and navigation. See Bilstein[15] for many more details on Warner's life and accomplishments.

Arthur Raymond, designer of the DC-3, was one of Warner's early students, receiving his master's degree in aeronautical engineering from MIT in 1921. Raymond and Warner remained lifelong friends, and much later, when Douglas was designing a new four-engine transport, the DC-4, Raymond called in his old MIT professor as a consultant. The influence of Warner and *Airplane Design: Aerodynamics* marched on.

With this, we march on – to the design of the Douglas DC-3, and its grand designer, Arthur Raymond.

[13] Ibid, p. 34. [14] Ibid, p. 34.

[15] Bilstein, Roger, E., "Edward Pearson Warner and the New Air Age," *Aviation's Golden Age*, William M. Leary, Ed., University of Iowa Press, Iowa City, 1989, pp. 113–127.

SETTING THE STANDARD FOR THE DESIGN REVOLUTION – THE DC-1 TO THE DC-3

The genesis of many airplane designs is competition. So it was in 1932, when Boeing was putting the final touches to the prototype of its 247 airplane – a pioneering, low-wing monoplane, all metal plane, with twin-engines wrapped in the new NACA low-drag cowling and with retractable landing gear. A three-view of the 247 is shown in Figure 4.6a and in a photograph in Figure 4.6b.

The Boeing 247 carried ten passengers in a soundproof cabin at speeds near 200 mph. The airplane was designed mainly by Charles N. Montieth, then Boeing's chief engineer, and Clairmont L. Egtvedt, the designer-in-charge. Montieth was a student of Edward Warner at MIT, receiving his M.S. degree from MIT in a special program for Army aviation cadets. He served as an Army lieutenant at McCook Field in Dayton where he was chief of the airplane section, and where he wrote one of the few respected early textbooks on applied aerodynamics.[16] He joined Boeing in 1925. Somewhat conservative in his design philosophy, he nevertheless shepherded Boeing's newly designed Model 200 Monomail and the revolutionary Boeing B-9 bomber into the age of the design revolution. Egtvedt joined Boeing immediately after graduating from the University of Washington in 1917, well before that university received the Guggenheim grant that would jump-start its Department of Aeronautical Engineering. He was to eventually become famous as the talented chairman of the Boeing Aircraft Company.

These gentlemen and their small design team produced the first modern airliner. The Boeing 247 was expected to revolutionize commercial air travel. Because of this, the airlines were standing in line for orders. However, Boeing at that time was a member of the United Aircraft Group, which included Pratt and Whitney Engines and United Airlines. Hence, United Airlines was first in line to receive the first seventy new 247s to come off the production line. This put the other airlines in an untenable competitive position.

Because of this, on August 5, 1932, Donald W. Douglas, president of Douglas Aircraft Corporation, received a letter from TWA. Dated August 2, the same letter had been sent to the Glenn Martin Company in

[16] Monteith, C. N., *Simple Aerodynamics and the Airplane*, Engineering Division, McCook Field, Dayton, Ohio, April 1924. Later revised with the aid of C. C. Carter at the United States Military Academy and published as *Simple Aerodynamics and the Airplane*, Montieth and Carter, The Ronald Press Company, 3rd Ed., 1929.

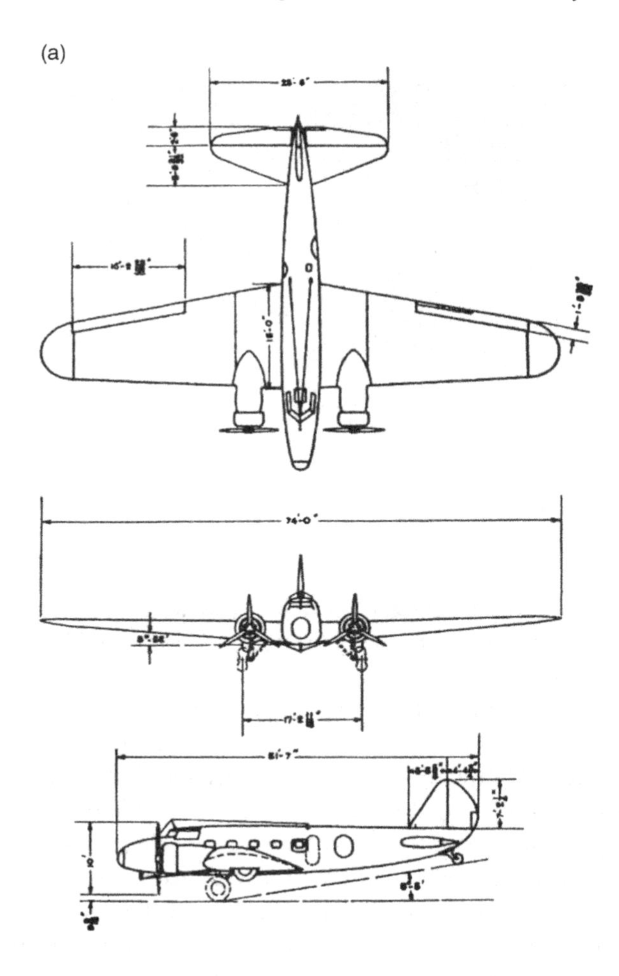

FIGURE 4.6. (a) Three-view of the Boeing 247D airliner, 1933.
(b) The Boeing 247D.
(National Air and Space Museum, SI-2006-29195.)

Baltimore, Maryland, and the Curtiss-Wright Corporation in St. Louis, Missouri, as well as to Douglas in Santa Monica. Jack Frye, a vice president of TWA, signed the letter. Frye was inquiring about Douglas's interest in designing a new commercial transport airplane. Because TWA could not readily obtain the new Boeing 247, they went after their own state-of-the-art airplane in an aggressive fashion. Attached to Frye's letter was a one page list of general performance specifications for the new airplane. (We note that the US Army's list of specifications that led to the

FIGURE 4.6. (*cont.*)

purchase of the Wright Military Flyer in 1908 was also one page long; clearly, twenty-five years later airplane specifications could still be given in a short, concise, clear-cut manner.)

The specifications called for an all-metal trimotor airplane that would have a cruising range of 1,080 miles at a cruise velocity of 150 mph. Of greatest importance, however, was the requirement that the new airplane at a full takeoff gross weight of 14,200 pounds be able to take off safely from any TWA airport with one engine out. At that time, the highest airport in the TWA system was in Winslow, Arizona, at an elevation of 4,500 feet. Other aspects of the specifications called for a maximum velocity of at least 185 mph, landing speed of not more than 65 mph, a minimum rate of climb at sea level of 1,200 feet per minute, and a minimum service ceiling of 21,000 feet, compromised downward to 10,000 feet with one engine out. (Recall that the stipulation of specifications is the first step of the intellectual methodology of conceptual airplane design set forth by Frank Barnwell.)

Donald Douglas and Jack Frye had met several times before, at various aviation functions in the Los Angeles, California area. They held a strong mutual respect for each other. Since the formation of his company in 1921, most of Douglas' business dealt with designing and constructing military airplanes, especially a successful line of torpedo airplanes for the Navy. However, he had been recently thinking about venturing into the commercial airplane market as well. (Airline passenger service had skyrocketed since Charles Lindbergh's historic solo flight across the Atlantic Ocean in 1927.) So Douglas paid serious attention to Frye's letter. He took it home with him that night, staying awake until 2 am pondering the ramifications. The next day he met with his core engineering design group and went over the TWA specifications one by one. One member of this group was Arthur Raymond, soon to become famous as the chief designer of the DC-3. The discussion lasted well into the evening. The group had already made the decision to submit a proposal to TWA; Friday's discussion was to be about the basic nature of the airplane design itself.

The TWA specifications called for a "trimotored monoplane" as being preferred, but held out the possibility of the design's being a biplane. Trimotor monoplanes were not new; the Fokker F-10 and the Ford trimotor had been flying in airplane service for almost five years. However, this airplane configuration suffered a public setback on March 31, 1931, when a TWA Fokker trimotor crashed in a Kansas wheat field, killing among the passengers the famous Notre Dame University football coach Knute Rockne. As for the biplane configuration, the early 1930s was a period of drag reduction via streamlining, and biplanes with their higher drag were on the way out.

So when the Friday meeting started out, it was no surprise that the chief engineer, James H. "Dutch" Kindelberger, stated emphatically:

> I think that we're damn fools if we don't shoot for a twin-engined job instead of a trimotor. People are skeptical about the trimotors after the Rockne thing. Why build anything that even looks like a Fokker or Ford? Both Pratt and Whitney and Wright-Aeronautical have some new engines on the test blocks that will be available by the time we're ready for them. Lots of horses ... any two of them will pull more power than any trimotor flying right now.

Douglas agreed. An essential design decision was made without making a single calculation.

Arthur Raymond, Kindelberger's assistant, who had earned a master's degree in aeronautical engineering at Massachusetts Institute of Technology in 1921 (one of the few people with graduate degrees in aeronautical engineering at that time), was immediately thinking about the wing design. He suggested, "Why not use a modified version of Jack Northrop's taper wing? Its airfoil characteristics are good. The taper and slight sweepback will give us some latitude with the center of gravity." Raymond was referring to the innovative wing design by Jack Northrop, who had worked for Douglas between 1923 and 1927 and then left for

Lockheed, finally forming his own company in 1931. (This is the same company that today builds the B-2 stealth bomber.) Northrop had developed a special cantilever wing that derived exceptional strength from a series of individual aluminum sections fastened together to form a multicellular structure (Figure 4.5). The wing is the heart of an airplane, and Raymond's thinking was immediately focused on it. He also wanted to place the wing low enough on the fuselage that the wing spars would not cut through the passenger cabin (as was the case with the Boeing 247). Such a structurally strong wing offered some other advantages. The engine mounts could be projected ahead of the wing leading edge, placing the engines and propellers far enough forward to obtain some aerodynamic advantage from the propeller slip stream blowing over the wing, without causing the wing to twist. Also, the decision was to design the airplane with a retractable landing gear. In that regard Douglas said, "The Boeing's got one. We'd better plan on it too. It should cut down on the drag by 20 percent." Kindelberger then suggested "Just make the nacelles bigger. Then we can hide the wheels in the nacelles." The strong wing design could handle the weight of both the engines and the landing gear.

The early 1930s was a period when airplane designers were becoming appreciative of the advantages of streamlining to reduce aerodynamic drag – one aspect of the design revolution. Retracting the landing gear was part of streamlining. Another aspect was the radial engines. Fred Stoneman, another of Douglas' talented designers, added to the discussion: "If we wrap the engines themselves in the new NACA cowlings, taking advantage of the streamlining, it should give us a big gain in top speed." This referred to the research at the NACA Langley Memorial Aeronautical Laboratory, beginning in 1928, that rapidly led to the NACA cowling, a shroud wrapped around the cylinders of air-cooled radial engines engineered to reduce drag greatly and to increase the cooling of the engine. At this stage of the conversation, Ed Burton, another senior design engineer, voiced a concern: "The way we're talking, it sounds like we are designing a racing plane. What about this 65 miles per hour landing speed Frye wants?" This problem was immediately addressed by yet another senior design engineer, Fred Herman, who expressed this opinion: "The way I see it, we're going to have to come up with some kind of an air brake, maybe a flap deal that will increase the wing area during the critical landing moment and slow the plane down ... Conversely, it will give us more lift on takeoff, help tote that big payload."

The deliberation extended into days. However, after a week of give-and-take discussions, they all agreed the airplane design would:

1. Be for a low-wing monoplane.
2. Use a modified version of the Northrop wing.
3. Be for a twin-engine airplane, not a trimotor.
4. Have retractable landing gear, retracted into the engine nacelles.
5. Have some type of flaps.
6. Use the NACA cowling.
7. Locate the engine nacelles relative to the wing leading edge at the optimum position as established by some recent NACA research.

The design methodology and philosophy exemplified by these early discussions between Douglas and his senior design engineers followed a familiar pattern. *No new, untried technology was being suggested.* All of the design features itemized were not new. However, the *combination* of all seven items in the *same* airplane design was new. Once again, we see a case where the new design did not involve a change in the basic methodology of conceptual airplane design, but rather rested on *technology* hung on the skeleton of Barnwell's methodology. The Douglas engineers were looking at past airplanes and past developments and were building on these to scope out a new design. To a certain extent, they were building on the Northrop Alpha (Figure 4.7). Although the Alpha was quite a different airplane (single-engine transport carrying six passengers inside the fuselage with an open cockpit for the pilot), it also embodied the Northrop multicellular cantilevered wing and an NACA cowling. Also, it was not lost on Douglas that TWA had been operating Northrop airplanes with great success with low maintenance.

During this first critical phase of their deliberation, the small team of Douglas designers had progressed through an intellectual process that reflected the impact of the maturing technology on the discipline of airplane design. In addition, they used more than a slide rule for calculations; they drew also on the collective intuitive feelings of the group, honed by experience. They practiced the *art* of airplane design to the extreme. At the end of the week, a proposal to TWA was prepared, and Arthur Raymond and Harry Wetzel (Douglas's vice president and general manager) took a long train ride across the country to deliver their proposal to the TWA executive office in New York. There, a three-week series of intense discussions took place; among the TWA representatives

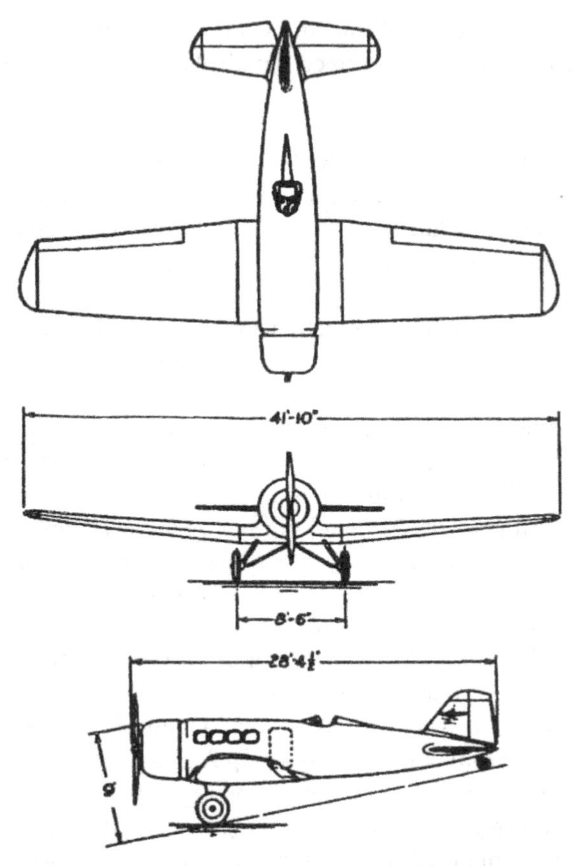

FIGURE 4.7. Three-view of the Northrop Alpha, 1932.

present at many of these meetings were Richard Robbins (president of TWA), Jack Frye, and Charles Lindbergh (the same Charges Lindbergh who had gained fame for his transatlantic solo flight in 1927 and who served as a technical consultant to TWA).

Although there were other competitors for the TWA contract, Raymond and Wetzel were successful in convincing TWA of the merits of a twin-engine ("bimotor") airplane over a trimotor. A major aspect of this consideration was the ability of the airplane to fly successfully on one engine, especially to take off at full gross weight from any airport along the TWA route and to be able to climb and maintain level flight over the highest mountains along the route. This was not a trivial consideration, and Raymond's calculations had a certain degree of uncertainty – the uncertainty that naturally is associated with the early aspects of the conceptual design process period. Raymond called from New York to tell Douglas about the critical nature of the one-engine-out performance's being a pivotal aspect of the discussions with TWA. When Douglas asked Raymond about his latest feelings as to whether the airplane design could meet this performance requirement, Raymond's reply was "I did some slide-rule estimates. It comes out 90 percent yes and 10 percent no. The 10 percent is keeping me awake at night. One thing is sure, it's never been done before with an aircraft in the weight class we're talking about." Douglas conferred with Kindelberger, who took the stand: "There's only one way to find out. Build the thing and try it." Douglas made the decision – Raymond should tell TWA that the Douglas Company would be able to construct such an airplane.

On September 20, 1932, in Robbin's office, the contract was signed between TWA and Douglas to build the airplane. Douglas christened the project as the DC-1, the Douglas Commercial One. The contract called for the purchase by TWA of one service test airplane at the cost of $125 thousand with the option (indeed, clear intent) of purchasing up to sixty additional airplanes, in lots of 10, 15, or 20 at $58 thousand each. The contract was forty-two typewritten pages long, twenty-nine of which dealt specifically with the technical specifications. There was a detailed breakdown of the empty weight and an expanded five-page list of performance specifications. The contract even went to the detailed extent of specifying such items as this: "Air sickness container holders shall be located adjacent to each seat in such a position as to be easily reached with seat in any adjustment." (Actually, this was not as trivial as it may seem today; the airplane was unpressurized, and hence, it would be flying, as did all aircraft at that time, at low altitudes where there was plenty of air turbulence, especially in bad weather.)

The concern that the Douglas designers put into the aspect of one-engine-out flight is reflected in a detailed technical paper written by Donald Douglas and presented by Douglas at the Twenty-Third Wilbur Wright Memorial Lecture of the Royal Aeronautical Society in London on May 30, 1935. The annual Wilbur Wright Lectures were (and still are) the most prestigious lectures of any society. It was a testimonial to Douglas' high reputation that he had received the society's invitation. The paper was entitled "The Development and Reality of the Modern Multi-Engine Airplanes after Engine Failure." Douglas began his paper with a statement that is as apropos today as it was then: "Four essential features are generally required of any form of transportation: speed, safety, comfort, and economy," However, today we would add *environmentally clean* to the list. Douglas went on:[17]

> The airplane must compete with other forms of transportation and with other airplanes. The greater speed of aircraft travel justifies a certain increase in cost. The newer transport planes are comparable with, if not superior to, other means of transportation. Safety is of special importance and improvement in this direction demands the airplane designer's best efforts.

Douglas then concentrated on engine failure as it related to airplane safety. He wrote:

> Statistics show that the foremost cause of accident is still the forced landing. The multi-engine, capable of flying with one or more engines not operating, is the direct answer to the dangers of an engine failure. It is quite apparent, however, that for an airplane that is not capable of flying with one engine dead the risk increases with the number of engines installed. Hence, from the standpoint of forced landings, it is not desirable that an airplane be multi-engine unless it can maintain altitude over any portion of the air line with at least one engine dead. Furthermore, the risk increases with the number of remaining engines needed to maintain the required altitude. In general, therefore, the greatest safety is obtained from
>
> 1. The largest number of engines that can be cut out without the ceiling of the airplane falling below a required value.
> 2. The smallest number of engines on which the airplane can maintain this given altitude.
>
> For airplanes equipped with from one to four engines, it follows that the order of safety is according to the list following.

Douglas followed with a list of ten options, starting with the category "four-engine airplane requiring one engine to maintain given altitude" as the most safe and "four-engine airplane requiring four engines to maintain good altitude" as obviously the least safe. Fourth down on the list was the two-engine airplane requiring one engine to maintain given altitude – this was the category of the DC-1 (and the DC-2 and DC-3 to follow). It is statistically safer than a three-engine airplane requiring two engines to maintain given altitude, which was fifth on Douglas's list.

[17] Douglas, Donald W., "The Developments and Reliability of the Modern Multi-Engine Air Liner with Special Reference to Multi-Engine Airplanes after Engine Failure," *The Aeronautical Journal*, Vol. 39, November 1935, pp. 1,010–1,042; also reprinted in *Journal of the Aeronautical Sciences*, Vol. 2, no. 4, July 1935, pp. 128–152.

Another hallmark governed the early design of the DC-1, namely creature comfort. This was particularly emphasized by Art Raymond who, after the TWA contract negotiations were over in New York, chose to fly back to Santa Monica. Flying from coast to coast at that time was an endurance test, especially in the Ford trimotor that Raymond was on. Raymond suffered from the noise, vibration, cold temperature at altitude, small and primitive lavatory facilities, uncomfortable seats, and even mud splashed on his feet. Indeed, he complained later, "When the airplane landed on the puddle-splotched runway, a spray of mud, sucked in by the cabin air vents, splattered everybody." After returning to the Douglas plant, Raymond stated, "We've got to build comfort, and put wings on it. Our big problem is far more than just building a satisfactory performing transport airplane." The team set about immediately to design an airplane that included soundproofing, cabin temperature control, improved plumbing, and no mud baths.

In 1932, the Guggenheim Aeronautical Laboratory at the California Institute of Technology (GALCIT) had a new, large subsonic wind tunnel. It was the right facility in the right place at the right time. Situated at the heart of the southern California aeronautical industry at the time when that industry was set for rapid growth in the 1930s, the California Institute of Technology (Caltech) wind tunnel performed tests on airplane models for a variety of companies that had no such testing facilities. Douglas was no exception. As conceptual design of the DC-1 progressed into the detailed design stage, wind-tunnel tests on a scale model of the DC-1 were carried out in the Cal wind tunnel. Over the course of 200 wind-tunnel tests, the following important characteristics of the airplane were found:

1. The use of a split flap increased the maximum lift coefficient by 35 percent and increased the drag by 300 percent. Both effects are favorable for landing.
2. The addition of a fillet between the wing and the fuselage increased the maximum velocity by 17 mph.
3. During the design process, the weight of the airplane increased, and the center of gravity shifted rearward. For that case, the wind-tunnel tests showed the airplane to be longitudinally unstable. The design solution was to add sweepback to the outer wing panels, hence shifting the aerodynamic center sufficiently rearward to achieve stability. The mildly swept-back wings of the DC-1 (also used on the DC-2 and DC-3 airplanes) gave these airplanes enhanced aesthetic beauty as well as a distinguishing configuration.

Wind-tunnel testing played an important role in fine-tuning the design of the DC-1, hence of the DC-2 and DC-3 to follow. What a change compared to the attitude of most airplane designers in the previous decades, who eschewed the wind tunnel in their design process. Frank Barnwell, who relied on aerodynamic data from wind tunnels at the National Physical Laboratory, and who managed to convince Bristol Aircraft to build their own wind tunnel at the end of World War I, was an exception. By the late 1920s, however, at the beginning of the design revolution, wind-tunnel data had become more reliable, and airplane designers began to use wind tunnels to fine-tune their designs. Indeed, witness the statement by E. P. Warner in his 1927 quote at the beginning of this chapter that the modern student at that time was inundated with *experimental data*, i.e., wind-tunnel and flight data. In regard to the DC-1, Dr. W. Bailey Oswald – at that time a professor at Cal tech who was hired by Douglas as a consultant on the DC-1 aerodynamics – later said, "If the wind-tunnel tests had not been made, it is very possible that the airplane would have been unstable, because all the previous engineering estimates and normal investigations had indicated that the original arrangement was satisfactory." Beginning in 1928, Arthur Raymond taught a class on the practical aspects of airplane design at Caltech, and Oswald attended the class in the first year. The two became trusted colleagues. Finally, reflecting on his first action on returning to Santa Monica after his trip to the TWA offices in New York, Raymond later wrote:[18]

> The first thing I did when I got back was to contact Ozzie (Oswald) and ask him to come to Santa Monica to help us, for that one-engine-out case still bothered me. I told him we only needed him for a little while, but he stayed until retirement in 1959, and ultimately had a large section working for him.

On July 1, 1933, the prototype DC-1 was ready for its first flight. It took less than one year from the day the original TWA letter arrived in Douglas's office to the day that the DC-1 was ready to fly. The story of that first fight is told at the beginning of this chapter. At the end of its test flight program, the airplane had met all its flight specifications, including the one-engine-out performance at the highest altitudes encountered

[18] Raymond, Arthur E., "Recollections of Douglas 1925–1960; From the DC-3 to the DC-8," *Journal of the American Aviation Historical Society*, Vol. 32, No. 2, Summer, 1987, p. 114.

along the TWA routes. It was a wonderful example of successful, enlightened airplane design.

An interesting contrast can be made in regard to the time from design conception to the first flight. During World War I, some airplanes were designed by laying out chalk markings on the floor and rolling out the finished airplane two weeks later. Fifteen years later, the process was still relatively quick, that for the DC-1 being about eleven months. Compare this to the design time for today's modern civil and military airplanes, which sometimes takes close to a decade between design conception and first flight.

Only one DC-1 was built. The production version, which involved lengthening the fuselage by 2 feet and adding two more seats to make it a 14-passenger airplane, was labeled the DC-2. The first DC-2 was delivered to TWA on May 14, 1934. Altogether, Douglas manufactured 156 DC-2s in twenty different models, and the airplane was used by airlines around the world. It set new standards for comfort and speed in commercial air travel. But the airplane that really made such travel an *economic success* for the airlines was the next outgrowth of the DC-2, namely, the DC-3.

As in the case of the DC-2, the DC-3 was a result of an airline initiative, not a company initiative. Once again, the requirements for a new airplane were being set by the customer. This time the airline was American Airlines, and the principal force behind the idea was a tall, soft-spoken, but determined Texan, Cyrus R. Smith. C. R. Smith had become president of American Airlines on May 3, 1934. American Airlines was operating sleeper service, using older Curtiss Condor biplanes outfitted with Pullman-sized bunks. On one flight of this airplane during the summer of 1934, Smith, accompanied by his chief engineer, Bill Littlewood, almost subconsciously remarked, "Bill, what we need is a DC-2 sleeper plane." Littlewood said that he thought it could be done. Smith lost no time. He called Douglas to ask if the DC-2 could be made into a sleeper airplane. Douglas was not very receptive to the idea. Indeed, the company was barely able to keep up with its orders for the DC-2. Smith, however, would not take no for an answer. The long-distance call went on for two hours, costing Smith more than $300. Finally, after Smith virtually promised that American Airlines would buy twenty of the sleeper airplanes, Douglas reluctantly agreed to embark on a design study. Smith's problem was that he had just committed American Airlines to a multimillion-dollar order for a new airplane that was just in the imagination of a few men at that time, and the airline did not have that kind of money. However, Smith then

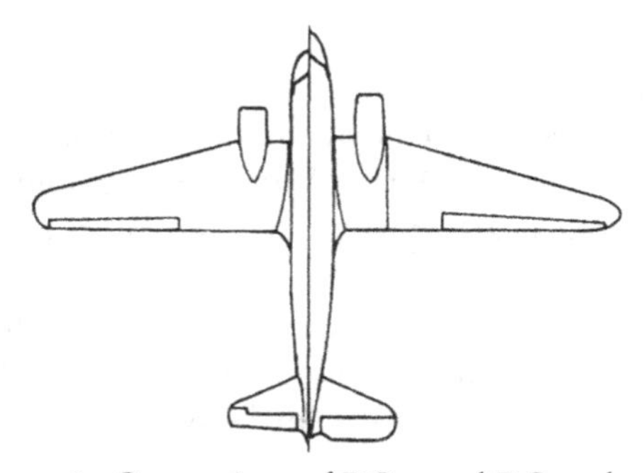

FIGURE 4.8. Comparison of DC-2 and DC-3 plan views.

traveled to Washington, D.C. to visit his friend and fellow Texan Jesse Jones, who was the head of Reconstruction Finance Corporation, a New Deal agency set up by President Franklin Roosevelt to help US business. Smith got his money – a $4.5 million loan from the government. The new project, The Douglas Sleeper Transport (DST), was on its way.

Design work on the DST, which was quickly to evolve into the DC-3, started in earnest in the fall of 1934. Once again, model tests from the Cal tech wind tunnel were indispensable. The new design outwardly looked like a DC-2. But the fuselage had been widened and lengthened, the wingspan increased, and the shape of the rudder and vertical stabilizer were different. In the words of Arthur Raymond, now the chief designer for the DC-3 project, "From the DC-1 to the DC-2, the changes were minor: from the DC-2 to the DC-3, they amounted to a new airplane." The different plan view shapes of the DC-2 and DC-3 are shown in Figure 4.8. The wind-tunnel tests at Caltech were overseen by Professor A. L. Klein and Bailey Oswald. During the tests, a major stability problem was encountered. Klein stated, "The bigger plane with its change in the center of gravity had produced the stability of a drunk trying to walk a straight line." However, by slightly modifying the wing and changing the airfoil section, the airplane was made stable; indeed, the DST finally proved to be one of the most stable airplanes in existence at that time. The first flight of the DST was on December 17, 1935. After the efforts of more than 400 engineers and drafters led by Arthur Raymond, the creation of 3,500 drawings, and some 300 wind-tunnel tests, the airplane flew beautifully. American Airlines began service of the DST on June 25, 1936.

The distinguishing aspects of the DST compared with the DC-2 were that its payload was one-third greater and its gross weight was about

50 percent larger. These aspects did not go unappreciated by Douglas. If the bunks were taken out and replaced by seats, the airplane could carry twenty-one passengers in a relative state of luxury. This was yet another new airplane – the DC-3. In fact, by the time Douglas gave his 1935 annual report to his board of directors, the DC-3 was already moving down the production line in parallel with the DST.

Less than 100 airplanes in the sleeper configuration, DST, were produced. But when the DC-3 production line was finally shut down at the end of World War II, 10,296 had been built. The vast majority of these were for the military, 10,123 compared with 803 for the commercial airlines. The DC-3 was an amazing success and today it is heralded by many aviation enthusiasts as the most famous airplane of its era.

A three-view of the DC-3 is given in Figure 4.9a. Compare this with the three-views of the Bristol F.2A and F.2B in Figure3.1. What a difference! This comparison dramatically illustrates the design revolution that ushered in the era of the mature propeller-driven airplane. Only nineteen years separate these two airplanes, but they are eons apart technically. Concentrating on Figures 4.9a and b, let us examine the advances in aeronautical engineering technology that were synergistically embodied in the DC-3 and that thereafter characterized all airplanes of the period.

The most obvious difference is the streamlined appearance of the DC-3, one of the most aesthetically beautiful of all aircraft. In fact, the configuration of the DC-3 screams *drag reduction*. This is one of the main aspects of the design revolution. Specifically, the drag-reduction features of the DC-3 are as follows:

1. *Streamlining of the basic airframe*. There is a smooth, elongated fuselage, somewhat round in cross section, and a wing with a relatively new airfoil shape generated by the National Advisory Committee for Aeronautics (NACA), designed to produce reasonable lift with little drag. Aerodynamicists had made progress in the theoretical understanding of the physical mechanisms that cause drag, and this understanding was beginning to feed into the design of airplanes.
2. *High aspect ratio wing*. The wing of the DC-3 had an aspect ratio of 9.14, considerably larger than its closest competitor, the Boeing 247, with an aspect ratio of 6.55, and certainly larger than the aspect ratio of 6 of the Bristol F.2B, as can be seen from a comparison of Figures 3.1, 4.6 and 4.9. By the time the DC-3 was designed, the aerodynamic theory of wings had advanced to the point that the

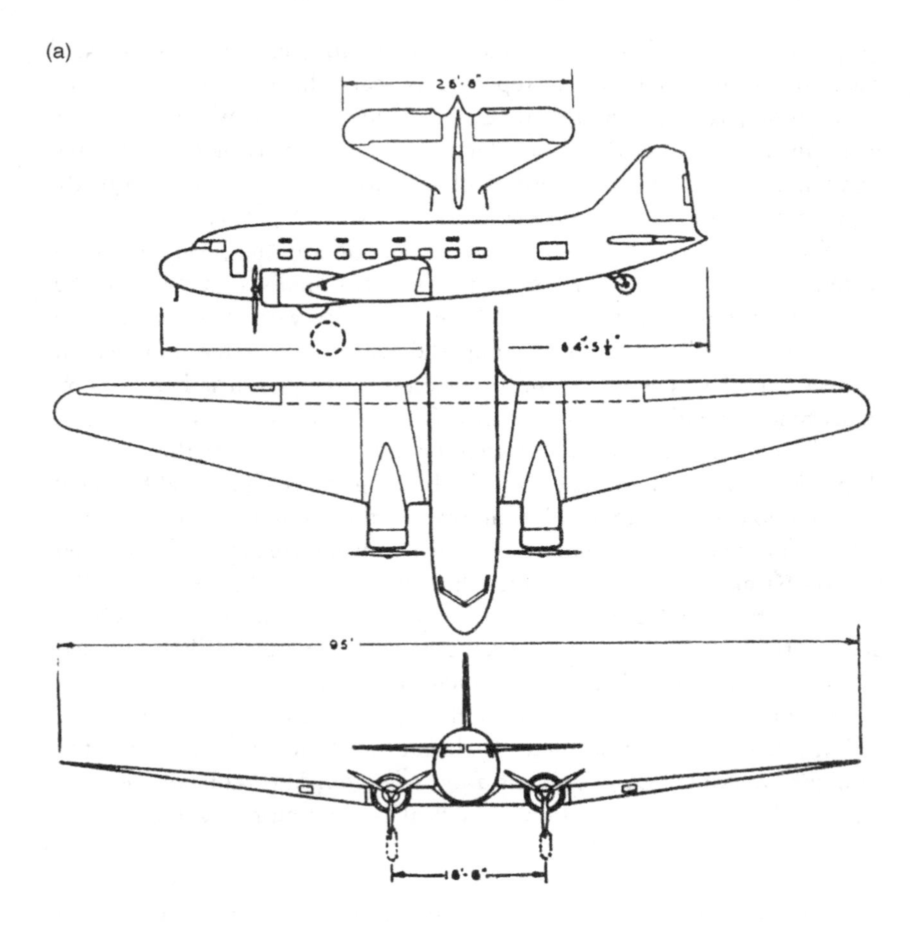

FIGURE 4.9. (a) Three-view of the DC-3.
(b) Douglas DC-3.
(National Air and Space Museum, NASM-9A05043.)

mysteries of aspect ratio were fully revealed, and accurate mathematical equations were available to predict the effect of aspect ratio on both lift and drag.

3. *NACA cowlings on the engines.* The DC-3 was powered by two Wright Cyclone engines – air-cooled radial engines. In a look at Figure 4.9, however, the cylinders of the engines are nowhere to be seen; they are covered over by streamlined shrouds, cowlings, specially designed by NACA based on a concerted wind-tunnel test

(b)

FIGURE 4.9. (*cont.*)

program that began in 1928. The story of the development of the NACA cowling was discussed earlier in this chapter. In some cases, the addition of the NACA cowling to a fuselage with engine cylinders previously exposed to the airstream resulted in a dramatic 60 percent reduction in drag.

4. *Retractable landing gear*. The fixed landing gear constantly protruding into the airstream and causing high drag, such as seen in Figure 3.1 for the Bristol fighter, and characteristic of virtually all airplanes from the era of the strut-and-wire biplane fighter, is missing in Figures 4.6 and 4.9. Indeed, one of the hallmarks of the era of the mature propeller-driven airplane is that the landing gear is retracted after takeoff and is neatly buried in the fuselage, wing, or nacelle. It is simply no longer hanging out in the airstream, thus reducing the parasite drag of the airplane. Aeronautical engineers as early as World War I appreciated the aerodynamic advantages of retractable landing gear, but the

> extra weight and mechanical complexity of the retracting mechanism worked against its implementation. Finally, in the early 1930s, the attention to drag reduction had brought about such dividends that the fixed landing gear was the next thing to go. For the DC-3, as well as several of the pioneering aircraft of the new era, the wheels were not completely retracted out of the airstream. Note in Figure 4.9 that the landing gear retracted into the engine nacelle, and because of competition for space in the nacelle between the engine and the wheels, the landing gear does not fit flush inside the nacelle. A small portion of the wheel is still exposed to the flow, but this is a small sacrifice to pay for the overall drag reduction obtained by getting most of the bulky landing gear out of the airstream. By the end of this era, most airplane designs had fully retracted landing gear, with a totally flush external airframe.

All of these features contributed to the streamlined airplane. Streamlining became a cardinal design principle during the design revolution; drag reduction was paramount.

Although you cannot easily tell from looking at the three-view in Figure 4.9a, the DC-3 was equipped with *variable-pitch propellers* manufactured by the Hamilton Standard Company. In contrast, the Bristol fighter, along with all airplanes of that era, had a fixed-pitch propeller. Before the 1930s, this was a weak link in all propeller-driven aircraft. The first practical variable-pitch propeller was not available until 1933, just in time for its use on a later model of the Boeing 247 and on the DC-1, followed by the DC-2 and DC-3.

Return to Figure 4.9, the only evidence that the DC-1 has a variable-pitch propeller is the small hub that projects slightly forward of the center of the propeller blades. This hub houses part of the hydromechanical control mechanism for varying the pitch of the propeller. The improved efficiency, and hence increased thrust, provided by the variable-pitch propeller allowed the DC-1 to meet the single-engine, high-altitude performance requirements set down by TWA. The variable-pitch propeller is one of the most essential ingredients of the mature propeller-driven airplane. It allows the airplane to, so to speak, shift gears in flight.

The DC-3 had a maximum speed of 212 mph, which made it the fastest commercial transport of its day. One design feature of the DC-3 that allowed this higher speed was its high wing loading of 25.3 pounds per square foot; by comparison, the wing loading of the slower Boeing

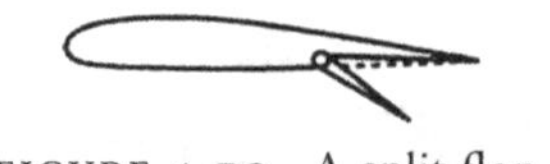

FIGURE 4.10. A split flap.

247 was much smaller, 16.3 pounds per square foot. Everything else being equal between two airplanes, the one with the higher wing loading will have a higher maximum velocity. A higher wing loading, however, comes with a price – a higher landing speed, unless something special is done about it. The Douglas engineers knew what they had to do to keep the stalling speed and hence the landing speed, a safe, low value. Because stalling speed varies inversely with the square root of the maximum lift coefficient, $C_{L,max}$, the wings of the DC-3 were designed with split flaps, which when deflected downward, artificially increased the value of $C_{L,max}$, and hence reduced the landing speed. The stalling speed of the DC-3 with its flaps deployed was a respectable 67 mph, only slightly faster than the 61 mph stalling speed for the Boeing 247, which was not equipped with flaps. A split flap is shown schematically in Figure 4.10; it forms part of the bottom surface of the wing and this bottom surface is the only part that deflects downward, whereas the top surface remains straight. This is in contrast to a plain flap, where the entire trailing edge is deflected. In the top view of the DC-3 shown in Figure 4.9, the flap-like surfaces shown at the trailing edges of the wings are long ailerons, not flaps. The dashed lines connecting the two ailerons, traversing the rest of the span of the wings, represent the split flaps that are hidden from view from the top. Although plain flaps had been used on a few airplanes as early as World War I, the DC-3 was the first major airplane in which flaps were an essential design feature. Moreover, the Douglas engineers chose to use split flaps, invented by Orville Wright in 1920, because the split flap produces a slightly higher $C_{L,max}$, and less of a change in pitching moment compared with a plain flap. Many of the fighter airplanes used during World War II were equipped with split flaps. The use of flaps of some type is a feature of the mature propeller-driven airplane.

Another feature that made the DC-3 a fast airplane for its time was the two powerful Wright Cyclones engines, each producing 1,100 horsepower. The Wright Cyclone was a 14-cylinder, air-cooled radial engine, with the cylinders arranged in two rows. After the demise of the rotary engines near the end of World War I, the basic radial engine experienced a hiatus until the late 1920s, when both Pratt

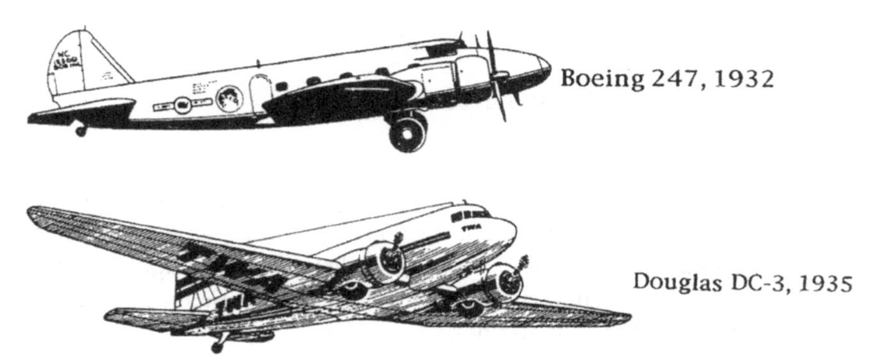

FIGURE 4.11. Boeing 247 and Douglas DC-3.

and Whitney Wasp engines and the Wright Cyclone engines, when combined with the low-drag NACA cowling, became the engines of choice of many airplane designers at the beginning of the era of the design revolution.

Again return to Figure 4.9. We mention an obvious feature of the DC-3 – it is a *monoplane*. Moreover, it is an *all-metal airplane*. The design features embraced by Hugo Junkers in 1915 had finally become the dominant airplane configuration by the early 1930s. For the most part, gone is the biplane configuration, gone is the vegetable airplane of wood and fabric, and gone is the thin airfoil. With the DC-3, we have an aluminum airplane with a cantilevered wing, made up of a 15 percent thick NACA 2215 airfoil shape at the root, tapering to a 6 percent thick NACA 2206 airfoil shape at the tip.

In summary, the Boeing 247 in 1932 and the Douglas DC-3 in 1935 (Figure 4.11) were products of the design revolution. With many technical features in common, they both introduced the modern airliner of the 1930s to a receptive and growing flying public. The DC-3, however, carried more passengers more comfortably and had somewhat better performance than the 247. It quickly became by far the preferred choice of the airlines.

In regard to the Douglas DC-3, we note in closing this section that the airplane itself embodied no new revolutionary design features per se. No new untried technology was used on the DC-3. Each aspect of the design revolution – such as the NACA cowling, wing flaps, retractable landing gear, etc. – had been tested and demonstrated separately. But what was revolutionary about the DC-3 was that it contained in its design, for the

FIGURE 4.12. Arthur Emmons Raymond.
(National Air and Space Museum, NASM-9A05043.)

first time, *all* of the features of the mature propeller-driven airplane and that Arthur Raymond combined all of these features in a synergistic fashion, resulting in one of the most technologically successful airplanes in history.

ARTHUR EMMONS RAYMOND: THE MAN

Arthur Raymond died on March 22, 1990 – just two days short of his 100th birthday. This aeronautical engineer (Figure 4.12), famous within professional circles as the designer of the Douglas DC-3 but virtually unknown to the general public, began his career seventy-three years earlier when he entered Harvard University and MIT in 1917. Four years later he graduated from Harvard with a B.A. and from MIT with a master's degree in aeronautical engineering. At MIT, he was one of Ed Warner's students in that early period when Warner was developing new courses in aeronautical engineering as well as generating the intellectual basis of his seminal book *Airplane Design*. Much later, Raymond would write about Warner:[19]

[19] Wright, p. 34.

He would continually multiply and divide numbers in his head in class while the class feverishly checked him on the slide rule and he would invariably be right to the third decimal. In spite of his youth, he had no difficulty in demonstrating his authority and superior knowledge in comparison with those he was teaching.

In 1921, the aircraft industry was in the postwar doldrums. Most remaining contracts for new airplanes were cancelled when World War I ended, and with the huge surplus of airplanes left over from the war, there was virtually no demand for new airplanes and new designs. Consequently, the industry had no demand for new aeronautical engineering graduates, even for graduates from MIT with a master's degree. Raymond, instead, went back to his roots. He was born in Boston on March 24, 1899. His father, Walter Raymond, who had founded the Raymond and Whitcomb Travel Agency a few years earlier, had built a 200-room resort hotel in Pasadena, California for his wealthy clients, and in 1903 moved his family to a cottage on the hotel grounds. Arthur grew up on these grounds and in 1921 returned to the Raymond Hotel to help his father and to wait for an opportunity in the aircraft industry. In the meantime, he married his high school sweetheart, Dorothy Lee, on June 30, 1921; their only son, Stanley, was born on May 22, 1922.

Arthur's mother, who was twenty years younger than his father, died unexpectedly after surgery in December of 1921, and although the rest of the family worked hard to accommodate the hotel guests, his father finally hired an experienced hotel man to assist him. Freed of some of the hotel duties, Raymond decided "the time had come for me to shift to my chosen field."[20]

In 1922, the Douglas Aircraft Company set up shop in an old motion picture studio on Wilshire Boulevard in Santa Monica, where it had some early success with the production of the famous *World Cruisers*, one of which resides in the Smithsonian's National Air and Space Museum. One day in early 1925, after watching the World Cruisers landing at Santa Monica's airport, Clover Field, Raymond decided to visit the factory to inquire about job openings. "I was told," Raymond later wrote, "that if a large contract for observation planes came in, expected in a couple of months, there might be a position in engineering – but in the meantime, would I take a position in the shop? I jumped at the chance and moved my family from Pasadena."[21]

Thus, in May of 1925 Arthur Raymond joined the Douglas Aircraft Company, where he started bending, filing, and drilling the corner

[20] Raymond, p. 110. [21] Ibid, p. 110.

connections for the cross-brace wires of tubular steel fuselages. During this period, Douglas called MIT looking for an engineer. Ed Warner told him that he already had a fine engineer working in his shop. Within six weeks, Raymond was sitting at a drafting table in the engineering department, not far away from the little cubbyhole where Donald Douglas sat with his drafting board, acting as both president and chief engineer.

Raymond recalled a typical interaction with Douglas at that time:[22]

> One day, Doug (as I was learning to call him) called me in and asked me if I could analyze the pontoon bracing for a Navy seaplane we were working on. I, manfully, said yes, then went back and tried to do it. I struggled for two days and nights without success before it came to me – just in time – that the truss had a redundant member. I had ways of handling that technicality in my calculations, and did so, just minutes before Doug called and asked for the answer. I gave it to him and all was well. If he had called earlier, my career might well have been nipped in the bud.

Arthur Raymond's master's degree from MIT had served him well.

Raymond's rising star paralleled the expansion of the Douglas Aircraft Company at that time. In the fall of 1925, J. H. "Dutch" Kindelberger left the Martin Company and joined Douglas as chief engineer. Kindelberger was an excellent engineer, designer, and executive. He had finished one year at the Carnegie Institute of Technology before World War I intervened. He enlisted in the Army Signal Corps, where he qualified as a pilot and flight instructor during the war. The differences in educational background made no difference; Kindelberger and Raymond quickly became close colleagues at work and their families became good friends at home. Also at this time, Jack Northrop occupied the drafting board next to Raymond, and Ed Heinemann and Jerry Vultee joined the company – names that would become icons in aeronautical engineering. With contracts for observation and cargo airplanes from the Army, torpedo planes from the Navy, and mail planes for Western Air Express, Douglas' yearly sales skyrocketed from half a million to two million dollars. Needing more space, the Douglas Aircraft Company moved to Clover Field in 1929. Engineering employment had doubled and was still increasing. And before the move Kindelberger had picked Raymond to be his assistant.

In 1928 the Guggenheim grant to Caltech touched Arthur Raymond. Dr. Robert Milikan, President of Caltech, wanted engineers from

[22] Ibid, p. 111.

industry to teach the new staff of the Guggenheim Aeronautical Laboratory at Caltech about the practical aspects of airplane design. Millikan approached Douglas for suggestions, and Douglas selected Raymond. Beginning in the fall of 1928, Saturdays found Raymond in the classroom teaching such well-known professors as Clark Millikan, son of Robert and head of the aeronautical engineering department; the Ph.D. physicist, Arthur Klein; Ernie Sechler; W. B. "Ozzie" Oswald; the theoretician Harry Bateman; and the self-taught experimentalist Albert Merrill. Raymond continued to teach part-time at Caltech until 1934 when Dutch Kindelberger left Douglas to become president of North American Aircraft, and Raymond succeeded to his job as chief engineer.

The previous section told the story of the design of the DC-1, DC-2, and DC-3; Arthur Raymond's role in this design is noted throughout that section. From the beginning, Dutch Kindelberger put Raymond in charge of the DC-1 project. The DC-1 design took place in a separate room off the main engineering department, highlighting the importance of this project to Douglas. The Douglas Aircraft Company had never designed a commercial, passenger-carrying transport; a lot was depending on the success of this project. Raymond had a small but expert team of designers working for him. Fred Herman was the project engineer, Ed Burton made most of the design layouts, and Lee Atwood did most of the stress analysis. Arthur Raymond orchestrated it all. When the DC-1 successfully flew on July 7, 1933, it put into the air the very embodiment of every major technical feature of the design revolution, all wrapped up in a single design. Raymond rather modestly wrote that "the DC-1 had the advantage of being able to combine a number of recent advances in the art."[23] What the DC-1 also had was the advantage of being designed by a grand designer.

There was only one DC-1 because it immediately morphed into the DC-2, of which 156 were built, and then evolved into the DC-3, arguably the most famous airliner in aeronautics history, of which 10,632 were built in the United States, and many more by the Russians in World War II. "From the DC-1 to the DC-2, the changes were minor," Raymond related; "from the DC-2 to the DC-3, they amounted to a new airplane. And the result was a nearly perfect fit to the market."

Art Raymond's passion for aeronautical engineering and flying began when, at the age of 15, he had a ride in a dirigible owned by Roy Knabenshue, at that time one of the leading proponents and designers of airships.

[23] Ibid, p. 114.

Raymond never looked back. His focus on aeronautics and his master's degree in aeronautical engineering placed him in a small select group of young engineers that blazed the trail to the design revolution in the 1930s. He had the drive to find an aeronautical engineering job in the 1920s when virtually no opportunities existed, and the resourcefulness to make the most of the opportunity when it presented itself. He was also fortunate to have that opportunity with Douglas because nowhere else would he have designed "the plane that changed the world," as some have called the DC-3. "In the DC-3, he did almost everything," his grandson Stephen eulogized. "He knew every bolt and screw in that airplane."

With the success of the DC-3, Douglas and the airlines began to look ahead to a larger airplane that ultimately became the four-engine DC-4. Once again, Raymond was in charge of the design. But unlike the DC-3 which was conceived by one airline and one company, the DC-4 was designed by a committee of five major airlines and Douglas. The original prototype had a triple tail, and flew in 1938. It did not measure up to expectations, however. The design had "too many cooks" according to Raymond. He took out a clean sheet of paper, made a fresh start on the design, and came up with a smaller and better DC-4 with a single tail. This airplane was a success, and during World War II, 1,200 were produced for the military under the designations C-54. During the Berlin airlift in 1948, many West Berliners became quite familiar with C-54s flying overhead, bringing in food and supplies, with some airmen dropping candy by small parachutes to anxious and waiting children below.

World War II brought a sea change in America's aircraft industry, creating massive expansion of existing factories and the creation of new plants. Douglas opened a major new plant in Long Beach and several smaller support facilities in Tulsa, Oklahoma City, and Chicago. Employment ballooned to 150,000 workers. With this, Art Raymond became a full-time executive. The chief engineers of each plant reported directly to him, and he was constantly traveling in order to maintain proper contact. Although he was consulted on various design problems, his career as a full-time airplane designer was essentially over.

In autumn of 1951, Raymond traveled to London to present the 39th Wilbur Wright Memorial Lecture at the Royal Aeronautical Society. He entitled his lecture "The Well-Tempered Aircraft," which Raymond defined as an airplane that was "well-conceived, well-executed, and well-shaken-down." Although not obvious from the title, Art Raymond used this occasion to record at one time and one place his philosophy on the whole airplane design process, from the very beginning of conceptual design to the final detail design including the shop production tools. He stated in his design philosophy that "an aircraft designed in such a way as not to take reasonable advantage of the state-of-the-art is handicapped competitively; one designed too far in advance of it, with too much optimistic anticipation, runs into the danger inherent of all pioneering efforts."[24] What better insight into the mind of the grand designer of the DC-3 is there? Raymond is simply expressing the tightrope walked by himself and his colleagues to include in the design of the DC-3 just the right number of technical advancements spawned by the design revolution.

He also expressed his viewpoint on the impact of wind-tunnel testing on airplane design. We have emphasized that before 1930, airplane designers did not have much faith in wind-tunnel data due to its sometimes unreliability and inaccuracy at that time, but with improvements in wind tunnel design and operation, airplane designers during the design revolution used wind tunnels to fine-tune their designs. The design of the DC-3 is a case in point. By the end of World War II, this process was taken a step father when airplane designers began to rely on wind tunnels at the very beginning of the conceptual design process. Raymond, however, had a rather conservative view of this matter, as he expressed in this paper:[25]

> Flight test time is expensive, particularly on modern aircraft, and can be shortened by suitable preliminary research and ground or laboratory testing, including wind tunnel testing. One of the difficulties of model testing in recent years has been the time required to build models and test them. The high speed at which these models must be run in the tunnel, the resulting high forces upon them, and the nature of the information which must be obtained from them, have produced model construction so intricate and test programmers so long that it has been difficult to get tunnel results early enough in the design of the aircraft to allow design modifications to be made while the drawings are fluid. In these circumstances, the wind tunnel tends to become a checking device rather than one for the collection of design information.

Raymond continued to reinforce his point:

> The tunnel tests may look questionable but unless the curves are very definitely out of line, the temptation is to wait and see whether the aircraft exhibits these characteristics in flight, rather than to interrupt the design schedule and delay the project to make changes which may not prove to be necessary. If there were

[24] Raymond, Arthur E., "The Well-Tempered Aircraft," *Journal of the Royal Aeronautical Society*, Vol. 55, Oct. 1951, p. 599.

[25] Ibid, p. 614.

complete confidence in the validity of wind-tunnel tests this would not be so, but too often there is doubt as to the correlation between these tests and flight. The result is that flight test merely confirms the fact that the wind-tunnel tests are right; the engineers congratulate each other that they knew it all the time, and the aircraft comes back to the shop for a major change.

This rather conservative view of the role of wind-tunnel testing in the design process, spoken in 1951, indicates a generational gap; Raymond's thoughts on this matter were still planted in the 1930s.

Such a view had more than just a philosophical impact. As Raymond was delivering the Wilbur Wright Lecture in London, Douglas engineers were carrying out the preliminary design work for a jet transport back in California – the DC-8. This was the beginning of a head-to-head competition between Boeing and Douglas to design and produce an effective commercial jet transport for the airlines, and the wind tunnel was to play a critical and defining role in the outcome.

In contrast to the conservative views expressed by Raymond about the role of wind-tunnel testing in the conceptual design process, Boeing practiced the exact opposite. In 1941, the company began the design of a new, large, low-speed wind tunnel. Boeing called in Caltech professor Theodore von Karman as a consultant, who recommended very strongly to Eddie Allen, head of Boeing flight test and aerodynamics, that the tunnel be designed for very high speeds. near the speed of sound. Von Karman was convinced that high-speed flight was just around the corner, although at that time the conventional wisdom argued that such a tunnel was not needed. Boeing was beginning to fine tune the design of their B-29 bomber, for which lower speed wind tunnels were quite adequate. Many in Boeing felt that the von Karman's recommendation was unsupportable and that the added expense of designing the tunnel for high speeds (which would about quadruple its cost) was unaffordable. Eddie Allen nevertheless prevailed, and Boeing went forward with the high-speed tunnel. Designed completely "in-house" by Boeing engineers with the help of two consultants, von Karman and MIT professor John Markham, the tunnel became operational in February 1944. In the test section, Mach 0.87 was achieved. The timing could not have been better. Fully checked out by the end of World War II, the tunnel was ready to test Boeing's new swept-wing XB-47 bomber. These were not tests to fine-tune the design but rather fundamental tests to explore and define the swept-wing configuration right at the beginning of the conceptual design process. William H. Cook, then a recent graduate of MIT with a master's degree, was in charge of the design and construction of the wind tunnel. "During the postwar years," Cook later wrote, "Boeing was uniquely fortunate among U.S. airplane manufacturers in having at its disposal the uninterrupted use of a high-speed wind tunnel, at a time when technical innovations were badly needed."[26] As chief of the technical staff in the transport division of Boeing, Cook played a role in the design of the Boeing 367–80, the prototype of the 707 jetliner. In reflecting on the competition between Boeing and Douglas for the first American commercial jet passenger airplane, and ultimately Boeing's success in the competition, Cook simply states that "what made the difference for Boeing was the wind tunnel."[27,28] He chided the Douglas management for not paying attention to a few engineers at Douglas who felt that the jet age was imminent, and that Douglas needed their own high-speed wind tunnel. In effect, Cook was chiding Arthur Raymond, and even pointed to Raymond's lecture on "The Well Tempered Airplane" as revealing how far Douglas lagged behind Boeing in wind-tunnel testing.

So we see that Arthur Raymond, the grand designer of the DC-3, was, like all the grand designers treated in this book, fallible. As vice president of Douglas, Raymond certainly played a role in the design of the DC-8, but it was much more of a business role rather than that of a "grand designer." Ivan Shogran, the longtime head of the power plant section, was the chief project engineer on the DC-8 and he reported to Raymond. Douglas was slow out of the gate with the DC-8 design. There was considerable pressure from the airlines, particularly United and American, to consider a turboprop transport rather than one powered by turbojets. Also, Douglas responded to intensive pressure exerted by C. R. Smith of American Airlines in late 1951 to design a more powerful version of its very successful DC-6 so that American could compete with TWA's new Lockheed Super Constellation. Designated the DC-7, this project absorbed considerable engineering and financial resources at Douglas, materially delaying the DC-8. Finally, in June of 1955, Donald Douglas decided to "bite the bullet," as Raymond related, and to go ahead with the DC-8 on company money. Thus began a rather hectic five-year period in Raymond's life where he faced major technical and business problems associated with the design of the DC-8. "The DC-8 was indeed the culminating experience of my

[26] Cook, William H., *The Road to the 707*, TYC Publishing Company, Bellevue, Washington, 199, 131.
[27] Ibid, p. 232.
[28] Raymond, *Recollection of Douglas 1925–1960; from the DC-3 to the DC-8*, p. 1,124.

career" he stated. "I don't regret it, but I wouldn't like to go through the experience again!"

Arthur Raymond retired from Douglas in 1960. He had accomplished much more in his career than being the grand designer of the DC-3. He was a pioneering founder of the RAND Corporation in 1945, the beginning of the "think tank" generation. In 1946, he was appointed by President Truman to a ten-year term on the National Advisory Committee for Aeronautics (NACA). He was a charter member of the National Academy of Engineering. These are the types of honors and accolades that go to a very successful member of the profession. But it all started when Art Raymond designed the DC-3.

What makes a grand designer grand? This is one of the questions we will be exploring in this book. Is it education? Frank Barnwell wrote the book establishing the basic methodology of conceptual airplane design. Arthur Raymond had a master's degree in aeronautical engineering from MIT and shared his knowledge with students and faculty at Caltech. Is it opportunity? Frank Barnwell had the opportunity to design airplanes for the Bristol Aeroplane Company for virtually all his professional career. Arthur Raymond had the opportunity to design airplanes for the Douglas Aircraft Company, spending his entire career with Douglas. What I think is most important, however, is focus, drive, and an exceptional ambition to design flying machines. Barnwell had it and so did Raymond. Perhaps there is no better indication of Raymond's focus and dedication to his work than that after his boyhood home, the Raymond Hotel, was foreclosed by the bank and demolished in 1934 and after his father died of a stroke in the same year, leaving behind an almost worthless estate, and in spite of being deeply affected by all these events, Raymond wrote:[29] "But I felt fortunate to be building a career outside, instead of going down with a sinking ship." And considering the loss of 83 acres of prime real estate with the hotel in Pasadena that later would be worth millions, Raymond wrote: "This does not bother me; my career has been rewarding in itself – and that is enough." This total career dedication is the stuff of a grand designer.

There were a number of other grand designers that came out of the design revolution in the 1930s. We will highlight two of them in the next two chapters.

[29] Ibid, p. 232.

5

Design for Speed

R. J. Mitchell and the Spitfire

It would be presumptuous of me to assume that speed has reached its finality; continuous development will lead to further progress. Speeds which amaze us today will be commonplace tomorrow.

R. J. Mitchell, in an interview with the *Southern Daily Echo*

I don't want anything touched.

Mutt Summers, Supermarine test pilot

The scene: Eastleigh aerodrome, southern England. *The time*: the afternoon of March 5, 1936. *The characters*: Mutt Summers, chief test pilot for Supermarine Aviation, his backup test pilot, Jeffrey Quill, and a small group of Supermarine personnel. *The action*: Sitting on the grassy surface of the aerodrome is a prototype of a new, advanced fighter, exceptionally streamlined, and with a striking aesthetic beauty that sets it apart from any airplane flying at that time. Supermarine was part of the larger Vickers-Armstrong works, and the chairman of Vickers, Robert McLean had unilaterally named this aircraft the "Spitfire." Parts of the airplane are unpainted, and the rest has a dirty yellowish-green factory primer finish. It does, however, have the traditional red, white, and blue roundel insignia of the Royal Air Force painted on the sides of the fuselage, with the identification K5054 painted on the rear of the fuselage and on the tail (Figure 5.1). The roundel is also painted on the top and bottom of both wings, which are elliptically shaped and part of the aesthetic beauty of the airplane. At 4:35 pm, Summers starts the engine, which itself is an object of mechanical engineering beauty. Designed by Rolls-Royce, and designated the Merlin C, the 12-cylinder inline engine is capable of 1,000 horsepower. All this, however, is hidden from view, contained within the long, narrow streamlined nose cowling of the

FIGURE 5.1. The prototype Spitfire, K5054, Eastleigh Aerodrome, 1936. (National Air and Space Museum, NAM-A-2329.)

airplane. Summers taxis over to the far side of the aerodrome, and takes off in the face of a slight 35 degree crosswind. The dual-pitch two-bladed propeller is set for optimum takeoff, and in the words of Jeffrey Quill (Quill, Jeffrey, *Spitfire*, John Murray Publishers Ltd, 1983, pp. 71): "The aeroplane was airborne after a very short run and climbed away comfortably." This first test flight calls only for checks on the stalling characteristics and the related behavior of the flaps. By instruction from R. J. Mitchell, the airplane's designer, Summers keeps the landing gear in the full down position. Eight minutes later, after gently cruising around at an altitude of 3,000 feet, the airplane and pilot are safely back on the ground. Summers taxis back to the hangar where the small group of observers, including Mitchell, are standing. As they moved closer, gathering near the cockpit, Summers pulls off his helmet and with much authority in his voice says "I don't want anything touched." This statement is to later become part of the folklore, implying that the airplane was perfect. What Summers means instead is that he does not want anything changed until he takes the next flight, which is to take place the next day. Nevertheless, Summers, a very experienced test pilot, is obviously elated about the success of this first flight of the Spitfire.

And so is R. J. Mitchell. The smile on his face penetrates his normally reserved look. Mitchell already inwardly knows the airplane will be all that he designed it to be. What he does not know is how important the

Spitfire will be to his country's eventual victory in the Battle of Britain to play out four years later.

Two days after the Spitfire's first flight, Hitler's troops marched into the Rhineland.

The Supermarine Spitfire went on to become one of the most famous World War II airplanes. It was the best performing fighter that the Royal Air Force had during the Battle of Britain. By the end of the war, almost 23,000 Spitfire and Seafires (the naval version) were produced in fifty-four different versions. The success of this airplane was mainly due to its designer, R. J. Mitchell, who easily ranks among the "grand designers" in aeronautical history. This chapter tells his story and examines his role in the evolution of the intellectual methodology of conceptual airplane design as played out in the middle of the design revolution in the 1930s.

R. J. MITCHELL, THE MAN

Reginald Joseph Mitchell (Figure 5.2) was born on May 20, 1895, in Stoke-on-Trent, in the midst of England's pottery industrial region. His father, Herbert Mitchell, was a school headmaster in Langton who was trained as a teacher at York College. His mother, Eliza Jane, was the daughter of a master cooper of Langton. A few months after R. J.'s birth, the family

FIGURE 5.2. Reginald Joseph Mitchell (1895–1937)

moved into a comfortable house named Victoria Cottage in Normacot, away from the smoky industrial atmosphere of the Pottery Towns. Here, R. J. Mitchell grew up in comfortable surroundings surrounded by a lawn and garden large enough for Mitchell and his four siblings to play with some freedom. His father soon afterwards gave up teaching and became a master printer at the firm of Wood, Mitchell and Company, Ltd. of Hanley. With hard work, and with much artistic ability, he later became managing director and finally the sole owner until his death in 1933.

R. J. Mitchell's formal education was at the Queensberry Road Higher Elementary School in Normacot, and later the Hanley High School. By all accounts, he enjoyed his school days, and excelled at mathematics. In 1911, he left school to become an apprentice at the firm of Kerr, Stuart and Company, a locomotive engineering company in Fenton, at the center of Stoke-on-Trent. There he served a five-year apprenticeship, starting first in the grimy engine sheds, and then moving on to the drawing office. He took advantage of the opportunity to attend the Wedgwood Bursham Technical School at night, taking classes in mechanics, higher mathematics, and engineering drawing. Again excelling in mathematics, he was awarded a special prize by the Midland Counties Union. This was, however, the extent of Mitchell's formal education. He never attended college, and never earned a formal engineering degree.

Nevertheless, his five years at Kerr Stuart's solidified his interest in engineering. With this, and with his burning interest in airplanes, R. J. left Kerr Stuart in 1916 at the age of 21 and looked for another job. World War I was then in its second year and R. J. attempted twice to join the military, being turned down each time because his engineering skills were viewed as more valuable. The Supermarine Aviation Works at Woolston, Southampton, also considered R. J. valuable; in 1917 they offered him the position of personal assistant to Hubert Scott-Payne, owner and manager. R. J. was so elated that upon being offered the job, he did not even return home; instead, he telegraphed his father to send his personal items directly to Southampton. Gordon Mitchell later theorized about R. J. that, "No doubt his quiet manner, combined with an aggressive jaw which gave his face a determined expression, persuaded Scott-Payne to give the eager young man from Stoke-on-Trent a chance."[1,2,3]

[1] Quill, Jeffrey, *Spitfire*, John Murray Ltd., 1983, p. 71.

[2] Gordon Mitchell, R. J. Mitchell, *World-Famous Aircraft Designer: Schooldays to Spitfire*, Nelson and Saunders, Olney, Buckinghamshire, England, 1986, p. 21.

[3] Ibid, p. 27.

The Supermarine Aviation Works was founded in 1912 by the Honorable Noel Pemberton Billing, adventurer, right wing politician, prolific inventor, theatrical impresario, magazine publisher, and aeronautical pioneer.[4] Energetic and eccentric, he dabbled in many projects but was unsuccessful in most of them. In his lifetime he took out over 2,000 patents, including one for a self-lighting cigarette. The straightlaced, monocle-wearing Billing spent much of World War I railing against what he perceived as the sexual decadence of the period, and leveling charges against the internationally famous exotic dancer, Maud Allen. He was an independent Member of Parliament for East Hertfordshire from 1916 to 1921, when he left due to poor health. His most important legacy, Supermarine, was conceived while he was living on a boat on the River Itchen in Southampton; he decided to establish his own aircraft company factory on the river bank. The works started out with one shed, a disused coalhouse, and some empty cottages.[5] Billings was enamored with flying boats and seaplanes, and these were the primary types of aircraft built by Supermarine for almost two decades. Indeed, the name of the company, "Supermarine," was chosen to describe boats that flew *above* water in contrast to submarines, boats that moved under water. In 1916, when he became an MP, Billing passed his control of Supermarine to his close friend Hubert Scott-Paine, a large man who liked speed – fast airplanes and motor boats. An ex-boxer, Scott-Paine was unorthodox, and perhaps that is why he hired R. J. Mitchell essentially right off the street, seeing in the young man an unusual enthusiasm and a drive to succeed. It was a prescient decision; over the next twenty years, Mitchell was to become Supermarine's most important asset.

An assistant to Scott-Paine, R. J. became familiar with the day-to-day activities of all the projects in the factory. The change from locomotives to airplanes was a natural match for him; both were complex machines designed for speed, requiring the ability to read detailed plans and complex blueprints. In 1918 he was promoted to assistant to the works manager, and in 1919 was made chief designer. A year later he became chief engineer of Supermarine.

MITCHELL'S DESIGN GROWTH

What Mitchell learned about airplane design in his early days with Supermarine, he learned by self-study and from the limited experience he

[4] McKinstry, Leo, *Spitfire. Portrait of a Legend*: John Murray, London, 2007, pp. 17–18.
[5] Gordon Mitchell, p. 25.

already had with the company. Unlike Art Raymond, who joined Douglas with a master's degree in aeronautical engineering from MIT, Mitchell joined Supermarine with no formal education in aeronautical engineering. However, the overpowering love of airplanes that dominated his life, and his ability for self-education, made up for any lack of formal education. For example, available to him in the early 1920s was Frank Barnwell's book and published papers on airplane design, setting the intellectual methodology for conceptual airplane design. The evening courses in mechanics and higher mathematics that he took at the Wedgewood Burslem Technical School gave him enough background to read the fledgling literature on airplane aerodynamics. In 1920, the seminal book *Applied Aerodynamics*[6] by Leonard Bairstow was published. The content of this book was based on Bairstow's experience at the National Physical Laboratory where he had been working since 1904, rising to become head of airplane research during World War I. At the time of publication of his book, Bairstow was appointed professor of aerodynamics at the Imperial College, becoming head of the aeronautics department there in 1923. Beginning in the 1920s, anybody who was into the technical aspects of aeronautical engineering in Britain read Bairstow's book. Indeed, his book was still being read well into World War II. This was the kind of technical literature available to R. J. Mitchell when he was just beginning his design career at Supermarine, and it had to be part of his design maturation. His intense desire to learn is noted in his obituary written by C. G. Grey, editor of *The Aeroplane*.[7] He wrote "One of the chief reasons for Mitchell's success was his willingness to learn." In a lecture to the Royal Aeronautical Society on January 21, 1954, given by Joseph Smith,[8] Mitchell's successor as chief designer at Supermarine, Mitchell is described as a man who "was always open-minded and took a great interest in aeronautical developments." In 1929, Mitchell was made a Fellow of the Royal Aeronautical Society, the premier organization for the dissemination of aeronautical engineering research in Britain through its series of lectures and its archive publication, *Journal of the Royal Aeronautical Society*. What Mitchell lacked in formal aeronautical engineering education, he more than made up in dedicated self-study.

[6] Bairstow, Leonard, *Applied Aerodynamics*, Longmans, Green and Company, 1920.

[7] Obituary for R. J. Mitchell, *The Aeroplane*, June 16, 1937, p. 718.

[8] Smith, Joseph, "R. J. Mitchell: Aircraft Designer," *Journal of the Royal Aeronautical Society*, May, 1954, Vol. 58, pp. 311–328.

FIGURE 5.3. Supermarine P.B. 31E Nighthawk, 1917.

Knowledge, per se, does not make a grand airplane designer; it is a necessary but not a sufficient condition. The ability to create, the almost intuitive instinct about what a new airplane should look like, and indeed even an aesthetic appreciation of the airplane, are qualities that come from within a person, and combined with knowledge can make that person a grand designer. R. J. Mitchell had all of these. These qualities were nurtured by the environment Mitchell found at Supermarine, a small company where the individual was prized and where a small number of engineers and technicians worked together on a large variety of aspects concerning a given airplane design and manufacture. When Mitchell joined Supermarine in 1917, the company was designing an aircraft for the purpose of intercepting German dirigibles that had been bombing Britain at night. Called appropriately the P.B. 31 Night Hawk, the airplane had four wings in order to provide the large wing area felt necessary to reach the altitudes at which the Germans were flying. The Night Hawk was an ungainly looking flying machine with little aerodynamic or redeeming aesthetic value (Figure 5.3), and its performance was disappointing, taking an hour to climb to 10,000 feet – definitely not an effective anti-Zepplin fighting machine. Mitchell's contribution to this airplane was minor, simply the preparation of some drawings of component parts. Only one Night Hawk was built, but this airplane was the beginning of his airplane-design learning curve.

By the time he was made chief designer in 1919, and chief engineer in 1920, Mitchell was already playing the major role in the design of new

FIGURE 5.4. Supermarine Southampton, 1925.
(National Air and Space Museum, NASM-7A40736.)

airplanes for Supermarine. For the next seven years, he worked to establish and enhance what became the bread-and-butter for Supermarine – the design and production of flying boats. The location of the Supermarine Works at Woolston on the River Itchen that fed into the Southampton Water was ideal for such production. The 1920s at Supermarine saw the design of no less than eleven successful flying boats and amphibians by Mitchell. All were biplanes, the standard feature for most airplanes of that time. Some were small (the single-engine Sea King weighing 2,850 pounds in 1921) and others were large (the twin-engine Swan weighing 12,832 pounds in 1924). Mitchell was following the standard design practice of basing each new design on the experience gained from the previous design. There were no design breakthroughs, and the incorporation of new technology was slow but steady. Gradually, more powerful engines became buried in streamlined nacelles, and the wing-tip floats used on some designs presented less frontal area and were connected to the wings by streamlined struts. Mitchell knew to pay homage to drag reduction, and he also played the role of aerodynamicist. It was not until 1931 that Supermarine hired an aerodynamic specialist, Beverly Shenstone, the first person with a degree in aeronautical engineering to join the company.

Mitchell's maturity in airplane design by 1925 is best illustrated in the Southampton flying boat (Figure 5.4). Designed for long-range reconnaissance, this military flying boat was an outgrowth of a previous Supermarine airplane, the Swan, but with a much more aerodynamically streamlined hull. It was considered at the time as "probably the most beautiful biplane flying boat that had ever been built, and certainly the most beautiful hull ever built."[9] Here we see one of the first clear examples of the aesthetic nature of Mitchell's design thinking. The role played by aesthetics in airplane design is subjective, and is still an area of research by the present author.[10] Mitchell clearly had artistic leanings from the time he was a young man. Mitchell's son Gordon calls attention to a hunting scene expertly painted by his father while still working for Kerr Stuart in 1916.[11] Delivering the first R. J. Mitchell Memorial Lecture before the Southampton Branch of the Royal Aeronautical Society on January 21, 1954, Joe Smith, Mitchell's successor as chief designer in 1937, reflected that Mitchell:

> ... was an inveterate drawer on drawings, particularly general arrangements. He would modify the lines of an aircraft with the softest pencil he could find, and then remodify over the top with progressively thicker lines, until one would finally be faced with a new outline of lines about three-sixteenths of an inch thick. But the results were always worthwhile, and the centre of the line was usually accepted when the thing was re-drawn.[12]

This "re-drawing" was based simply on Mitchell's intuition honed by experience and guided subjectively by his aesthetic nature. It was not based on any slide-rule calculations or technical formulas. This is perhaps why, as we shall soon see, Mitchell designed so many beautiful airplanes, pleasing to the eye.

According to Joe Smith, 1925 was the most successful year of R. J.'s career. The beautiful Southampton appeared and was labeled in the highly respected *Janes All the World's Aircraft* in 1926 as "one of the most notable successes in post-war aircraft design." A total of seventy-nine Southamptons were produced between 1925 and 1934, serving with distinction with Coastal Reconnaissance squadrons. Many of the later Southamptons had metal duralumin hulls rather than the wooden hulls of earlier models, an example of Mitchell incorporating new technology in

[9] Shelton, John, *Schneider Trophy to Spitfire*, Haynes Publishing, Sparkford, Yeovil, Somerset, UK, 2008, p. 92.

[10] Anderson, John D., Jr., "Design for Performance: The Role of Aesthetics in the Development of Aerospace Vehicles," in *Aerospace Design: Aircraft. Spacecraft. and the Art of Modern Flight*, edited by Anthony M. Springer, Merrell Publishers Limited, London, 2003, pp. 52–74.

[11] Gordon Mitchell, p. 22.

[12] Smith, Joseph, "R. J. Mitchell: Aircraft Designer," *The Aeroplane*, Jan. 29, 1954, pp. 125–127.

his designs. And 1925 saw the appearance of Mitchell's Supermarine S.4 Schneider Trophy Racer, a stunning new aircraft that was a complete departure from anything he had designed before, and a totally new experience for Supermarine.

MITCHELL'S SCHNEIDER TROPHY RACERS – NEW TECHNOLOGY PERSONIFIED

By 1912, the fledgling world of the airplane had become one of daredevil air shows and lucrative air races. On December 5, at a banquet of the Aero Club of France, Jacques Schneider announced his sponsorship of a new air racing trophy just for seaplanes in order to encourage their development. Schneider was a French financier, balloonist, and aircraft enthusiast. The Schneider Trophy is a rather large silver sculpture, almost risqué, of a winged figure kissing a zephyr reclining on a breaking wave. Schneider established the rule that the first country to win the Schneider race three times in a row would permanently retire and retain the trophy. The first Schneider Trophy race was held in Monaco in April 1913, but it was not until 1931 that the Trophy was retired, having been won by Britain. It is now in the collection of the British Science Museum in South Kensington, London, prominently displayed in their aeronautics gallery. The principal reason why it is there is due to the design genius of R. J. Mitchell.

The Schneider Cup races were interrupted by World War I and the third contest was not held until 1919, in Bournemouth, England. Mitchell by this time was in his third year of working for Supermarine and although he did not participate in its design, the company entered a modified version of its Sea Lion I, a single-engined biplane flying boat designed by F. J. Hargreaves, Mitchell's predecessor. The airplane crashed during the preparatory trials and Italy carried off the Trophy for that year. However, Mitchell and the company were hooked; after a three-year hiatus, a new Supermarine entry, the Sea Lion II, took off from the waters of the Bay of Naples in 1922 and won the sixth Schneider Trophy contest with an average speed of 145.7 mph. Although based on Supermarine's 1919 Schneider's entry, the Sea Lion II was an airplane redesigned by Mitchell. In subsequent publicity, the company described its success as "a very great tribute to the excellence of Supermarine design." Not willing to dabble much with success, Mitchell and Supermarine entered the 1923 Schneider Trophy competition held at Crowes, England with a slightly upgraded machine, the Sea Lion III, shown in Figure 5.5, sporting

FIGURE 5.5. Supermarine Sea Lion III, at the 1923 Schneider Trophy Race. (National Air and Space Museum, NASM-2A39889.)

an uprated engine in a somewhat streamlined nacelle, and new wing-tip floats with less frontal area (hence less drag) mounted on streamlined struts. But a glance at Figure 5.5 shows a still somewhat conventional single-engine biplane flying boat. The Americans, however, entered an airplane in the 1923 Schneider race that knocked the socks off everybody in attendance. The Curtis CR-3 racer was a complete departure from previous Schneider racer designs. Shown in Figure 5.6, it was not a flying boat but rather a highly streamlined float plane. This was a completely innovative new look in airplane design, and the American CR-3 won the seventh Schneider Trophy contest with an average speed of 177.38 mph; Mitchell's Sea Lion III placed third with an average speed of 157.17 mph.

This wake-up call was all that R. J. Mitchell needed. He recognized that the Sea Lion III embodied an obsolete design philosophy, and he quickly set out to design a completely new flying machine for Supermarine's next entry in the Schneider races. The result was the sleek, streamlined S.4 monoplane shown in Figure 5.7. It had a minimum of drag-producing protuberances, and the floats were state of the art, having been subcontracted to Shorts who had recently installed a testing tank for such items. The floats had low drag and a single step on the bottom to expedite takeoff (unsticking) from the water surface when the surface was smooth. The wing of the S.4 was unlike anything designed before by Supermarine, a cantilever structure with no external bracing wires. The radiators were thin and mounted on the bottom surface of the wings. The cooling water was carried back and forth to the engine through conduits

FIGURE 5.6. Curtiss CR-3 racer at the 1923 Schneider Trophy Race. (National Air and Space Museum, CW8G-T-2484.)

FIGURE 5.7. Supermarine S.4 Schneider Trophy racer, 1925. (National Air and Space Museum, NASM-1B42519.)

buried inside the wings. The wings and fuselage were covered with load-bearing plywood sheeting; no fabric covering was used. In a single bold stroke, Mitchell was incorporating new technology throughout the design of the airplane. But in no respect was the intellectual methodology used for S.4 conceptual design any different from that used by Mitchell in his earlier aircraft. Once again, the strikingly unique S.4 represented Mitchell's willingness to incorporate new technology within the framework of a tried and tested intellectual methodology for conceptual design.

Unfortunately, the S.4 suffered from wing flutter problems. With Mitchell watching, the S.4 stalled and crashed into the Chesapeake Bay on October 25, 1925, the day before the eighth Schneider Trophy race was held. Fortunately, the pilot, Henri Baird, survived, but the S.4 was out of the race. Lieutenant Jimmy Doolitle, flying a Curtiss R3C-2 biplane racer – now on permanent display in the Smithsonian's National Air and Space Museum in Washington, DC – won the event at an average speed of 232.5 mph. Mitchell returned to Southampton, greatly disappointed, but absolutely convinced that the monoplane racer configuration was still superior to the biplane. The S.4 was a revolution in airplane design in 1925 and it greatly influenced all subsequent Schneider racers. This is why Joe Smith in his R. J. Mitchell Memorial Lecture in 1954 considered 1925 the most successful year of Mitchell's career, with the design of two such disparate airplanes, the beautiful Southampton flying boat and the exceptionally streamlined S.4 racer.

Mitchell embracing new technology is personified in his design of the racing airplanes for subsequent Schneider Trophy races. Moreover, the Schneider competition, which heretofore had been sponsored and controlled by the aero clubs of the various competing countries, was now getting support from the various national governments. In Britain, the Air Ministry virtually demanded that wind-tunnel tests be carried out in support of future Schneider designs and in 1926 such testing was carried out in facilities at both the Royal Aeronautical Establishment (RAE) in Farnborough and the National Physical Laboratory (NPL) near London. At Supermarine, Mitchell was already considering his next Schneider airplane, the S.5. He was after three obvious design improvements: (1) weight reduction, (2) drag reduction, and (3) improved water performance. The latter is where Mitchell found the RAE and NPL wind-tunnel tests most useful, with the examination of ten different float configurations. In the design of the S.5, Mitchell incorporated:

(1) Copper radiators with flat surfaces mounted flush with the upper and lower wing surfaces.
(2) The addition of streamlined wire bracing between the wing, floats, and fuselage as a means to reduce the possibility of wing flutter. This also allowed a slight reduction in wing area which reduced friction drag and hence compensated for the extra parasite drag caused by the wires. Moreover, this reduced the weight of the wing in comparison to the purely cantilever structural design of the S.4.
(3) A lowering of the wing and a more forward location of the cockpit to improve the pilot's visibility compared to the S.4.
(4) Further reduction in the cross section area of the fuselage and in the frontal area of the floats, both for drag reduction.

FIGURE 5.8. Supermarine S.5 Schneider Trophy racer, 1927.
(National Air and Space Museum, NAM-A-38975-C.)

(5) A higher powered engine geared to improve propeller efficiency. The engine that Mitchell had chosen for the S.4 was a development of the proven Napier Lion and which produced about 700 horsepower. For the S.5, he went with a Napier Lion VII engine that produced 900 horsepower. Napier worked to reduce the frontal area of the engine, and along with closer cowling allowed the S.5 to have a more streamlined and smaller nose than the S.4.

All of these improved design features justified the new designation S.5 for Mitchell's new design.

In the 1920s, metal was beginning to replace wood as the designer's material of choice. Mitchell made the fuselage of the S.5 an oval, metal monocoque construction, keeping wood as the wing material with a skin of one-eighth plywood.

The Supermarine S.5 (Figure 5.8) was ready in 1927 for competition in the tenth Schneider Trophy Contest. On September 26, the S.5 streaked over the waters of the Adriatic Sea near Venice, winning the race with an average speed of 281.65 mph, a new record for seaplanes and exceeding the World Speed Record for landplanes by 3 mph.

By this time, the Schneider races were spaced two years apart. Flushed with success, Mitchell began to plan a yet modified racer for the 1929 event. It is in the psychology of airplane designers to base a new design on improvements made to the good aspects of a previous design. Here, Mitchell was no different. He was convinced of the correctness of his streamlined monoplane designs for the S.4 and S.5 and his next racer would exhibit the same features. However, he made one dramatic change – he changed engines. In 1927, Mitchell began to work personally

FIGURE 5.9. Supermarine S.6A Schneider Trophy racer, 1929, on display at the Solent Sky Museum, Southampton, England.

with Sir Henry Royce, one of the founders of the Rolls-Royce engine company. Rolls-Royce had been designing and producing aircraft engines since World War I, beginning with the very successful Eagle engine. Sir Henry took an immediate liking to Mitchell, and pledged the design of a new engine that would produce at least 1,500 horsepower for Mitchell's new racer. Given only six months to design, build, and test the new engine, the Rolls-Royce engineers at Derby came up with the soon to become famous 'R' engine, a very slimline power plant that would fit into Mitchell's new racer, the Supermarine S.6. Because of the increased overall size and weight of this new "super" engine, the S.6 design was slightly different than the S.5, requiring a different cowling, a slightly increased wing area, and a more forward location of the front float struts. For the most part, however, the S.6 (Figure 5.9) looked very similar to the previous S.5. But with the Rolls-Royce 'R' engine, it was really quite a different airplane.

On September 7, 1929, the S.6, powered with 1,850 horsepower from the 'R' engine, won the eleventh Schneider Trophy Contest at Calshot, England, with an average speed of 328.63 mph. Five days later, in a separate event, the S.6 reached 357.7 mph, establishing a new World Speed Record. This was the second Schneider Trophy race in a row won by Britain, and each one with a Supermarine racer designed by R. J. Mitchell. By this time, the racers, Supermarine, and Mitchell were receiving great press coverage and for the moment, Mitchell became almost a household name.

Mitchell, with no formal degree in aeronautical engineering and no initial pedigree in the formal engineering world, had by now become very well-known and respected in professional aeronautical circles, all on the strength of his self-made accomplishments in airplane design. He published little in the open literature, and did not encourage professional speaking engagements. However, in 1929 he was elected a Fellow of the Royal Aeronautical Society, bringing him into the fold of some of the most learned aeronautical engineers and scientists in England, and indeed throughout the world. The few papers written by Mitchell were short, direct, and succinct. Of particular relevance to this stage in his life was his paper in the December 25, 1929 issue of *Aeronautical Engineering*, a supplement to the periodical *The Aeroplane* edited by C. G. Grey. Titled "Racing Seaplanes and Their Influence on Design," this paper starts with Mitchell's personal assessment that "quite a lot of information and experience is gained in the development of racing aircraft which is of undoubted value to the designer in all branches of aeronautical engineering and in many ways this has had a pronounced influence on the design of both military and civil types of aircraft." Such a statement is still today somewhat controversial because there are many who feel that racing airplanes, because of their point design and very focused, specialized purpose have not contributed that much to the technical development of conventional military and civil aircraft. We will once again reflect on this statement when we later examine Mitchell's design of the Spitfire, and explore just how much of his experience with the Supermarine Schneider Trophy racers impacted the Spitfire design. In 1929, however, Mitchell put a premium on speed. In this paper, he states that: "The results obtained in the form of speed have been a direct and absolute indication of our progress in aeronautical development. Speed in the air," he writes, "must always be a measure of aerodynamic efficiency, which in turn must always be the most important consideration in all aircraft design." Much of the rest of the rather short (two page) paper is concerned with details of seaplane racer design, such as floats. In regard to the Rolls-Royce 'R' engine, however, Mitchell made a prescient statement that, "There is little doubt that this intensive engine development will have a very pronounced effect on our aircraft during the next few years." Within five years, the specialized 'R' engine would morph into the famous Rolls-Royce Merlin engine that helped to make the future Spitfire and other British airplanes so successful.

Britain was one win away from retiring the Schneider Trophy. The next race was scheduled for 1931. In a stunning reversal of its previous support, however, the government decided not to provide monetary or staff support for the next race, citing the worsening economic crises spreading worldwide. A rationale was given that enough data on high-speed flight had been obtained from the earlier races, and that the original purpose of Jacques Schneider in pioneering high-speed sea-planes had been met. In lieu of funding from the Air Ministry, private individuals came forth with individual donations to fund the next Supermarine racer, the most pivotal being 100,000 pounds from Lady Houston. With this, Mitchell and Supermarine proceeded to the next racer design.

Once again, not wishing to dabble with success, Mitchell used the airplane designer's favorite conceptual tool of basing the new design on the previous one. Indeed, for the next racer, Mitchell carried over most of the previous S.6 design but chose new floats with additional radiator surface, raising the total radiator surface area on the airplane to 470 square feet, about 50 percent of the total available exposed surface area for the entire airplane. Mitchell, in a later radio broadcast, likened the new airplane to a "flying radiator." In a nod to some government support of the new racer, albeit minimal, new wind-tunnel tests on the new larger floats were carried out at the National Physical Laboratory, leading to a more streamlined float with increased length and smaller frontal area. Also the Rolls-Royce engineers at Derby were increasing the available power of the 'R' engine; the tuned up engine was now producing a phenomenal 2,530 horsepower. These two aspects – improved float design and a more souped-up engine – were the only two major technology improvements made to the previous S.6 design. Not justifying a new designation, the new airplane was labeled simply the S.6B (Figure 5.10).

On September 13, 1931, the Supermarine S.6B streaked over the water at Calshot, averaging 340.08 mph, permanently winning the Schneider Trophy for Britain. It was, however, a somewhat empty victory, because that day the French and Italian entries were not able to compete due to previous accidents and mechanical problems. The S6.B had won the race by default, satisfying the rules by flying the course as prescribed. But its win was fully justified; like the previous winning Supermarine machines, Mitchell had designed not only high performance racers, but reliable ones that worked. On September 28, the S.6B went on to set a new world speed record of 407.5 mph.

Were Mitchell's Schneider Trophy racers a product of new advancements in the intellectual methodology of conceptual airplane design? In spite of the innovative nature of these airplanes, the answer is no! Rather, the success of the S.5. S.6, and S.6B racers was due to Mitchell's willingness to incorporate the best new technology in his designs. In so doing,

FIGURE 5.10. Supermarine S.6B Schneider Trophy racer, 1931. (National Air and Space Museum, NAM-A-4201-B.)

however, he was well on his way to becoming a "grand designer." (This author saw first-hand an example of Mitchell's use of new technology when examining the S.6B on display at the British Science Museum in London. The surface of the metal skin is absolutely smooth. Mitchell had used flush riveting for the construction of the racer. Flush riveting for airplanes was quite new at the time; it was strictly a drag-reduction feature. Many designers chose not to use flush riveting because it increased production costs. For example, the Douglas DC-3 (Chapter 4) used the more common round-head rivets that protruded into the flow above the surface. Mitchell continued the use of flush riveting when he designed the Spitfire.)

SPITFIRE PROGENITOR – THE TYPE 224

In 1931, the standard fighter in use by the RAF was the Bristol Bulldog (Figure 3.6) designed by Frank Barnwell. The Bulldog had a maximum speed of 174 mph at 10,000 feet; this was in striking contrast to the 407.5 mph speed of the Supermarine S.6B in the same year. Of course, this comparison is one of apples and oranges; the Bulldog was a military airplane designed for extended and reliable service, and the S.6B was a highly focused point design with one single purpose – to win an air race.

The S.6B won the Schneider Trophy on September 13, 1931. Just eighteen days later, the British Air Ministry published Specification F7/30 for the design of a modern land-based fighter "to be on par with any Air Force's equipment by the year 1940." The specification called for an all-metal aircraft with a maximum speed of at least 195 mph, but with the door left open to the highest speed possible beyond that. The Air Ministry was looking for a major improvement in British fighter airplanes, and encouraged innovation.

A copy of the specification arrived at the Supermarine factory in Woolston on November 5. Supermarine was not in the business of building fighter airplanes, but their spectacular success in designing the winning Schneider Trophy racers clearly put them in everybody's mind as a source for high-speed airplanes. Moreover, Supermarine's management and R. J. Mitchell were basking in the euphoria of having triumphed with their winning racing airplanes. One can argue that racing airplanes, including Mitchell's S racers, had a minimal impact on the *technology* of new but conventional aircraft. However, the *subjective* impact of racer designs on the thinking and intuition of aircraft designers was not minimal. By this time, Mitchell was confident of his ability to design high-speed airplanes. One hundred and six days after receiving the Air Ministry's F7/30 Specification, Supermarine tendered a design. This was the first step made by Supermaine that would four years later result in the successful flight test of the prototype Spitfire as related at the beginning of this chapter. If Mitchell's Schneider Trophy racers had never existed, it is highly unlikely that the Air Ministry would have invited Supermarine to participate in the F7/30 Specification, and even less likely that Mitchell would have submitted any design, let alone one that would ultimately result in the Spitfire. So it is relatively safe to say that the single most important impact of Mitchell's Schneider Trophy racers on the Spitfire was not enhanced technology, but rather the very existence of the Spitfire.

Mitchell's design for the F7/30 Specification was labeled the Supermarine Type 224. This airplane was *not* the Spitfire of the future; instead, the Type 224 was to cause Mitchell much grief, and ultimately was a failure. By appearance, it fell far short of the type of airplane that might have been expected from the designer of the Supermaine Schneider

FIGURE 5.11. Model of the Supermarine Type 224 fighter, 1934, on display at the Solent Sky Museum, Southampton, England.

winners. The 224 (Figure 5.11) was an almost clumsy looking gull-wing monoplane with fixed landing gear in large fairings. In 1928, Supermarine became part of Vickers Aviation, which in turn was part of the giant Vickers Armstrong armaments and ship building empire. Unlike Supermarine, Vickers had a wind tunnel. Mitchell had narrowed his design option to three airplanes, two with straight wings and one with a bent gull wing. Although Mitchell was never a strong proponent of the use of wind tunnels in the design process (unlike Frank Barnwell who had pressured Bristol into building a company wind tunnel, as described in Chapter 3), the use of the Vickers wind tunnel was the only way to make an intelligent choice of Mitchell's three designs. The wind-tunnel data clearly indicated the superiority of the gull-wing design. Indeed, Supermarine's written submission to the Air Ministry explicitly mentioned the impact of the wind-tunnel tests on the Type 224 design:

> Every effort has been made to keep this aircraft clean. Wind Tunnel tests have been carried out on a preliminary model and indicate that chassis drag and interference are appreciably reduced by bending the wing and reducing the chassis height.

The wind tunnel was also used to determine the best combination of anhedral for the inner portion of the wing, and dihedral for the outer portion. Moreover, Mitchell had chosen a rather thick airfoil for the wing, an RAF 34 airfoil shape for the outer wing section, changing to a symmetric 18 percent thick NACA 0018 airfoil at the fuselage. The use of a thick airfoil for the Type 224 was contrary to the use of a thin airfoil for the Schneider Trophy racers; Mitchell soon came to regret this choice of a thick airfoil.

Mitchell's increased dependency on wind-tunnel tests during his design process for the Type 224 was not consistent with his previous airplane designs, which used little or no wind-tunnel data. However, it was consistent with the trend of airplane designers during the design revolution in the 1930s, when the reliability of wind-tunnel data was improving, to use wind-tunnel testing for fine-tuning their designs. In this sense, Mitchell was exhibiting the use of new technology in his design.

The Supermarine submission to the Air Ministry included a table summarizing the predicted performance of the new design, where the prediction was entirely based on the wind-tunnel data.

	Required by	**Estimated from**
	Specification F7/30	Wind Tunnel Tests
Speed at 15,000 ft. (mph)	195	245.5
Landing speed (mph)	60	60
Service ceiling (ft.)	28,000	43,500
Time to climb to 15,000 ft. (minutes)	8.5	6.6

The predicted performance looked great! The Air Ministry issued a production contact for one Type 224 prototype.

The airplane first flew on February 19, 1934 with test pilot Mutt Summers at the controls. Its performance was disappointing, with a top speed of only 228 mph, well below the predicted speed of 246 mph. Moreover, it took a long time to climb to 15,000 feet, 9.5 minutes in comparison to the expected time of 6.6 minutes. No additional Type 224's were produced. In January 1935, the Air Ministry cancelled all further construction on the Type 224. The airplane that eventually came out of the F7/30 Specification was the Gloster Gladiator, a conventional fabric-covered biplane fighter that could muster a top speed of 242 mph and climb to 15,000 feet in 6.5 minutes. By the beginning of World War II in 1939, the Gladiator was obsolete.

Justifiably so, Mitchell was disappointed in the Type 224 performance. Wind-tunnel tests indicated much better performance. As a result, Mitchell became a wind-tunnel doubter, thinking that tests on small scale models produced unreliable data. The problem was well understood among aerodynamicists by that time. Skin-friction drag and flow separation (the cause of pressure drag) is governed by the Reynolds number, a dimensionless parameter defined as the product of flow velocity, air density, and vehicle size, divided by the viscosity coefficient of the air. In a fluid flow, the Reynolds number is proportional to the ratio of inertia force and friction force acting on an element of fluid in a flow. In a wind-tunnel test, the size of the model is generally an order of magnitude or smaller than the actual airplane, and hence the wind-tunnel data corresponds to a Reynolds number anywhere between ten to one hundred times smaller than the Reynolds number encountered in an actual flight. Therefore, the lift and drag coefficients measured for the model in the wind tunnel will not correspond to the lift and drag coefficients for the full-scale airplane in flight. In order to deal with this problem in the 1920s, the NACA at the Langley Memorial Aeronautical Laboratory in Hampton, Virginia, put a wind tunnel completely inside a tank pressurized to 20 atmospheres. It was called the Variable Density Tunnel; the air density inside was twenty times larger than standard sea level density, thus increasing the Reynolds number in the tunnel by a factor of twenty, much closer to that obtained in real flight. The NACA also took an alternate solution by constructing a very large wind tunnel with a 30 feet by 60 feet cross section test section, large enough that a full-size fighter airplane could be mounted in the tunnel. The tunnel, simply labeled the Full Scale Tunnel, went into operation in 1930 and remained in operation until 2010. Unfortunately, neither Vickers nor any other aeronautical facility in Britain had anything like these wind tunnels and airplane designers at that time had every right to be skeptical of using wind-tunnel data in their design process. Mitchell ultimately went to the extreme. No wind-tunnel data whatsoever was used to design the Supermarine Spitfire, with the single exception that the Spitfire used a standard NACA airfoil shape that was derived from tests carried out in the Variable Density Tunnel at Langley.

The failure of the Type 224 was a bitter pill for Supermarine and Mitchell to swallow, especially coming on the heels of the Schneider racer successes. With faith in his ability to design a state of the art fighter, Mitchell set out initially on a complete redesign of the Type 224; the new design had straight wings and a thinner airfoil shape. The Type 224 was powered by a Rolls-Royce Goshawk engine (an updated Kestrel S) with 535 hp. In an example of serendipity, Rolls-Royce was now working on a completely new engine capable of 1,000 hp; first known as the PV12 (for private venture 12-cylinder engine), it soon became the famous Merlin engine that powered numerous aircraft during World War II. Supermarine and Rolls-Royce, both initially working without contract funding, were on track for the design of a completely new flying machine powered by the Merlin. The Air Ministry quickly realized that Mitchell's new design, originally for reconsideration for Specification 7/30, was indeed a new airplane. Sir Robert McLean, Managing Director of Vickers

Aviation and Supermarine, christened this airplane the "Spitfire." (McLean had earlier pressured the Air Ministry to label the Type 224 as the "Spitfire," but this name really never caught on for the Type 224.) The Air Ministry had wanted for a long time a Mitchell-designed fighter, and in January 1935 it issued Specification F37/34 just to cover the new Mitchell design. The Spitfire was born.

THE SPITFIRE

Sir Robert McLean, long after his retirement from Vickers in 1938, published his version of how the Spitfire was conceived. In *The Sunday Times* on August 18, 1957, McLean wrote:

> I felt they (his design team) would do much better by devoting their qualities not to the official experimental fighter (Specification F7/30) but to a real killer fighter. After unfruitful discussions with the Air Ministry, my opposite number in Rolls-Royce, the late A. F. Sidgreaves, and I, decided that the two companies together should finance the building of such an aircraft.
>
> The Air Ministry was informed of this decision, and were told that in no circumstances would any technical member of the Air Ministry be consulted or allowed to interfere with the designer.

Leo McKinstry[13] found that the Vickers' archives and Air Ministry documents tell a different story. Vickers had its own design office, independent of Supermarine, and this office had designed a small, lightweight monoplane fighter named the Venom. It was state of the art in 1935, with an NACA cowling, retractable landing gear, and light Browning 0.303 machine guns in its straight, tapered wings. It used a three-bladed DeHavilland variable-pitch propeller. It weighed 1,000 pounds less than the Spitfire. However, it was not ordered in prototype form by the Air Ministry, and instead was carried on company funds by Vickers as a private venture. Jeffrey Quill, later to become the major test pilot for the Spitfire, called the Venom "a sporting little airplane," but it did not measure up to the Spitfire in performance.[14] The Venom was plagued with mechanical problems and its testing program ended with a final flight by Quill on February 3, 1938. Throughout its test program, the Venom remained a favorite of McLean's. Indeed, on July 26, 1937, in private correspondence to Vickers' director Sir Archibald Jameson, McLean wrote: "The Venom is a much smaller machine than the Spitfire,

[13] Leo McKinstry, pp. 50–51. [14] Jeffrey Quill, p. 108.

and would of course be much cheaper to build." He went on to write that the test pilots felt that the Venom was a better airplane than the Spitfire, a statement vigorously disputed by Quill.

Current scholarship indicates that at best, McLean was lukewarm about the Spitfire. In fact, much later Beverly Shenstone wrote in a letter in 1960:[15] "Without seeming to lack respect for Sir Robert McLean's ability and energy, in my opinion the Spitfire would not have been born if Mitchell had not been willing to stand up to McLean, particularly in the era when McLean quite clearly preferred the Venom concept to the Spitfire concept because it was cheaper and lighter." Here is another aspect of a "grand designer," a person who so firmly believes in the viability of his design and is willing to fight for it. As a result, Supermarine did indeed underwrite the preliminary design of the Spitfire during the last half of 1934. But Mitchell kept the Air Ministry fully informed of the progress of his design, and the Air Ministry in turn expressed much interest albeit not providing monetary support at that time.

The intellectual progress of Mitchell's design of the Spitfire can be seen in Figures. 5.12 a, b, c, and d. Three days after the first flight of the Type 224 on February 19, 1934, the disappointed Mitchell pulled out a clean sheet of paper and commenced to redesign the airplane. Out went the gull wing and fixed landing gear, as seen in Figure 5.12a. A thinner airfoil section was used, an NACA 2412 airfoil. Mitchell's selection of this airfoil section is another example of his use of advanced but proven technology in his designs. Earlier in 1934, the Supermarine aerodynamicist, Beverly Shenstone, had visited America where he was told about the new NACA airfoil design and testing program. This program generated the NACA "four-digit" family of airfoil shapes that had excellent aerodynamic behavior – relatively high lift and low drag. By September, Mitchell had improved his design with further streamlining of the fuselage – a cockpit fairing that extended all the way to the vertical tail (Figure 5.12b). Then, during the second or third week of November 1934, Mitchell made the decision to use an elliptically shaped wing – perhaps the single most distinctive aesthetic feature of the Spitfire.

The elliptic wing planform, however, was based on technical considerations, not on aesthetics. Aerodynamically, for a wing of given aspect

[15] McKinstry, p. 50.

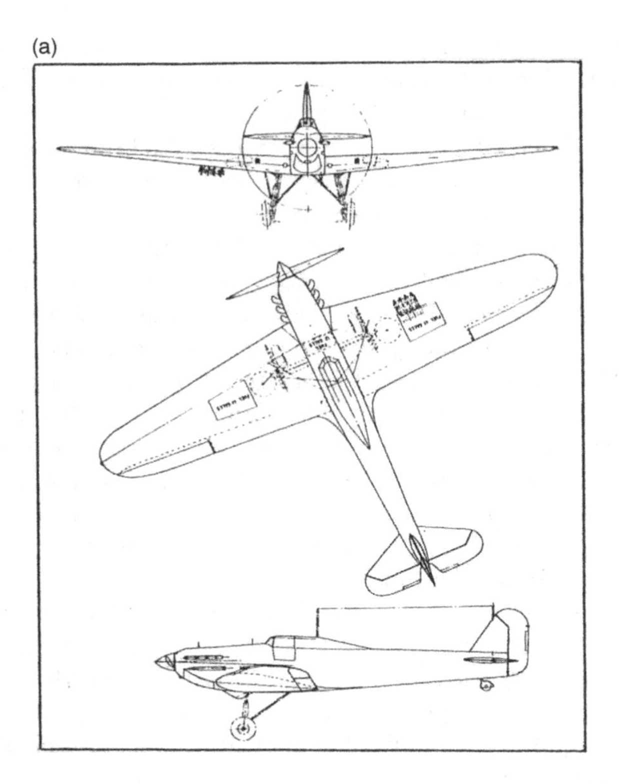

FIGURE 5.12. Evolution of Mitchell's conceptual design of the first Spitfire
(a) First version
(b) Second version
(c) Third version
(d) Final version
(e) The Supermarine Spitfire F.Mk 1.
(National Air and Space Museum, NAM-A-5067-1.)

ratio, the elliptic planform yields the lowest induced drag coefficient. This was first theoretically proven by the German fluid dynamicist, Ludwig Prandtl at Gottingen University, who during World War I developed his lifting-line theory for predicting the aerodynamic properties of a finite wing. (See Anderson, *A History of Aerodynamics*, Cambridge University Press, 1998). In 1921, Prandtl wrote a lengthy two-part technical report for the NACA detailing many of his advances in aerodynamic theory

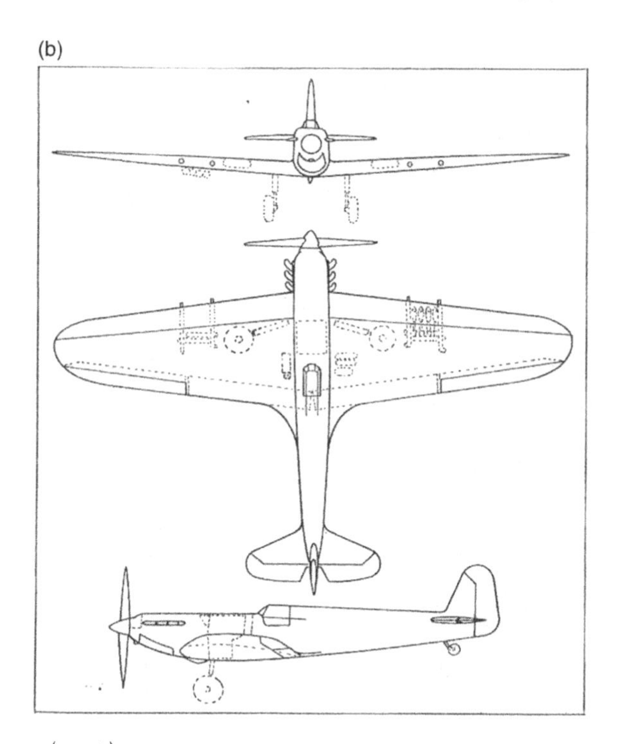

FIGURE 5.12. (*cont.*)

during the War.[16,17] That report was the first time that Prandtl's pioneering work appeared in English, and spreading Prandtl's work to the English speaking technical community was of course, the intent of the report. The benefit of an elliptical wing was clearly stated by Prandtl: "The elliptical distribution of lift, apart from its simplicity, has obtained a special meaning from the fact that the drag as calculated from equation (40) proved to be the smallest drag that is imaginable for a monoplane having given values of the total lift, the span and velocity." Later, in 1926 Herman Glauert, a well-known British aerodynamicist working for the Royal Aeronautical Establishment, published his seminal textbook, *The Elements of Aerofoil and Airscrew Theory*, that discussed at length

[16] Prandtl, Ludwig, *Applications of Modern Hydrodynamics to Aeronautics*, NACA TR 116, 1921.

[17] Glauert, Hermann, *The Elements of Airfoil and Airscrew Theory*, Cambridge University Press, 1926.

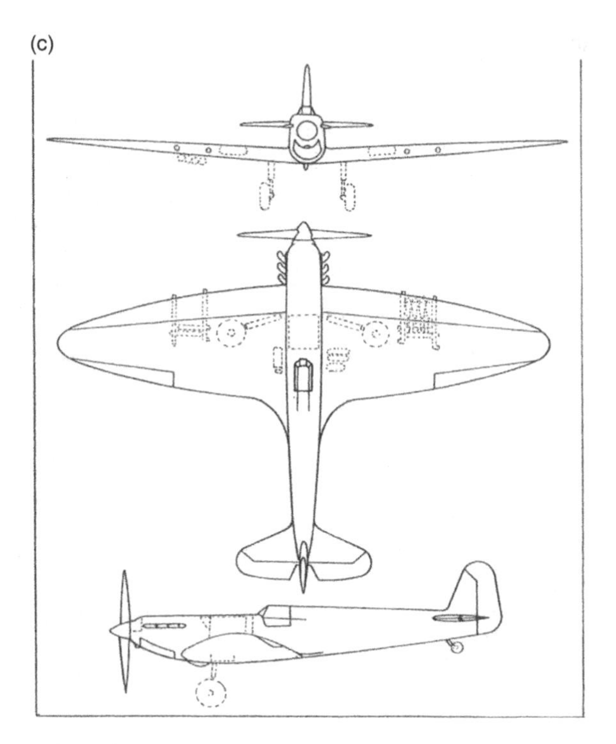

FIGURE 5.12. (*cont.*)

Prandtl's lifting-line theory. Clearly, in 1934, both R. J. Mitchell and Beverly Shenstone at Supermarine were well aware of the aerodynamic advantage of an elliptic wing.

Mitchell's decision in favor of the elliptic wing, however, was not based on aerodynamics. In 1934 the Air Ministry had moved towards the eight-gun fighter. Eight 0.303-inch machine guns were felt necessary to deliver enough ammunition on target to bring down an enemy bomber in the two-second burst thought available to the fighter pilot in combat. An elliptic wing had the advantage of a large chord length near the fuselage as well as geometrically sustaining a large chord for a reasonable distance away from the fuselage along the span. Mitchell saw this shape as providing plenty of room to fit four machine guns and ammunition within each wing, and leaving room to store the retracted landing gear as well. At first, Mitchell's design team presented him with the specific elliptic wing configuration shown in Figure 5.12c, which had the

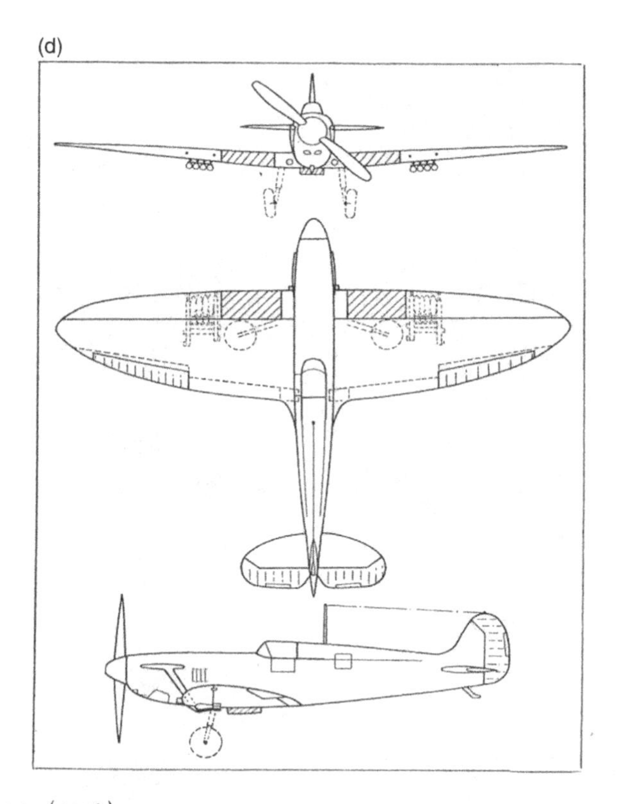

FIGURE 5.12. (*cont.*)

structural disadvantage of a bent spar along the span of the entire wing. The final configuration (Figure 5.12d) bent the wing forward in order to straighten the spar. In this sense the final wing planform was not quite a true ellipse, but the distribution of the chord length along the span followed an elliptical variation.

R. J. Mitchell as chief designer made the final decision on the Spitfire wing shape. His thinking is reflected in a comment he made to Shenstone: "I don't give a bugger whether it is elliptical or not, so long as it covers the guns."[18] Mitchell also wanted a thinner airfoil than the 18 percent thick airfoil used on the Type 224; he felt that choice to be a mistake. For the Spitfire, Mitchell and Shenstone chose a 13 percent thick NACA

[18] Shelton, p. 212.

(e)

FIGURE 5.12. (*cont.*)

2213 airfoil at the wing root, tapering to a 5 percent thick NACA 2205 airfoil at the wing tip. With a thin airfoil, the use of an elliptical wing with its large chord length near the root was particularly advantageous; the physical wing thickness was larger than it would have been for a straight tapered wing, with a shorter chord length, thus allowing enough height within the wing structure for the guns and landing gear. Mitchell's choice of a thinner airfoil was based mainly on obtaining lower drag. Little did he realize at the time that the thin Spitfire airfoil, markedly thinner than standard practice for World War II fighters, would also yield a serendipitously higher critical Mach number, thus allowing the Spitfire to achieve higher speeds in a vertical dive before adverse compressibility effects took hold. Much later, toward the end of the war, the critical Mach number for the Spitfire was measured at about 0.93, the highest of any contemporary fighter. In 1934, few aerodynamicists were aware of such compressibility problems. (One of the few was John Stack, an NACA aerodynamicist working in the 11 inch High-Speed Tunnel at the Langley Memorial Laboratory. In the January 1934 issue of the *Journal of the Aeronautical Sciences* published by the Institute of the Aeronautical Sciences (IAS) in the United States, Stack had a paper entitled "Compressibility Effects in Aerodynamics" detailing some of the NACAs latest research on the problem.[19] It was the first paper to point out that adverse compressibility effects would be encountered by high-speed airplanes in the future. Indeed, he even mentioned Mitchell's Supermarine S.6B as an example. There is no evidence that Mitchell read this paper while designing the Spitfire, although at the time the IAS journal would have been available to him at least in the library of the Royal Aeronautical Society in London, and Mitchell was a Fellow of the Society.)

Mitchell's comment that he did not give a "bugger" about the elliptical shape of the wing is curious, given that he knew that shape was optimum for minimum induced drag. I have made a calculation, detailed in Appendix A, of how much the maximum speed of the Mark I Spitfire would have been reduced if a straight tapered wing of the same aspect ratio had been used in place of the elliptical wing, i.e., if the configuration in Figure 5.12a had been used instead of that in Figure 5.12d. The result from Appendix A indicates a reduction of only 2.74 mph, only a 0.76 percent reduction below the 362 mph maximum velocity of the Spitfire Mark I. This result is not entirely surprising. In steady, level flight, the lift must always equal the weight of the airplane. At high speeds, much of the lift comes from the high dynamic pressure, which is proportional to the square of the velocity. In turn, the required lift coefficient (and the

[19] Stack, John, "Effects of Compressibility on High Speed Flight," *Journal of the Aeronautical Sciences*, Vol. 1, No. 1, Jan. 1934, pp. 40–43.

associated angle of attack of the airplane to the incoming relative wind) is small. The induced drag coefficient, which is proportional to the square of the lift coefficient, is therefore also small, and constitutes only a small fraction of the total drag coefficient of the airplane. The elliptical wing planform minimizes the induced drag coefficient, but since the induced drag is only a minor player at high speeds, the effect of the wing shape is not so important at high speeds. It is interesting, nevertheless, that the result calculated in Appendix A shows such a small effect of the elliptical wing planform on the maximum velocity of the Spitfire – less than 1 percent.

Mitchell's utterance that he did not care about the wing shape as long as it covered the guns came from one of the best airplane designers of that time. A "grand designer" does not make such a statement unless he knows the facts. Had Mitchell made a calculation that reflected the same kind of result as obtained in Appendix A? There is no record of any such calculation. Much more likely is that he simply *knew* by experience that induced drag for the airplane was small at high speeds, and hence the elliptical planform, although theoretically the shape that minimized induced drag, was not an important consideration. Much more important was each wing's capacity to hold four machine guns. For this practical requirement, the elliptic wing was advantageous on a purely mechanical rather than an aerodynamic basis.

In addition, the elliptical wing, along with the beautifully streamlined fuselage, certainly made the Spitfire one of the most aesthetically beautiful airplanes in history. As noted earlier, Mitchell appreciated artistic beauty, and painted some rather serous watercolors when he was in his early twenties. I have always felt that an appreciation for aesthetic beauty subjectively influenced some airplane designers. Later on in the life of the Spitfire, the elliptic wing posed some serious production difficulties and wing production lagged far behind the manufacture of the fuselage, delaying the introduction of the Spitfire into active RAF service by almost a year. Kenneth Agnew, professor of industrial design at the University of Ulster, writes:[20] "The problem of building these wings accurately was not solved until 1942." He goes on to state that "The Spitfire in production took more than twice as many man-hours to build compared to German fighters, and the number of Spitfires available during the crucial air battle of 1940 was less than half the total of Hawker Hurricanes (the other

[20] Agnew, Kenneth, "The Spitfire: Legend or History? An Argument for a New Research Culture in Design." *Journal of Design History*, Vol. 6, No. 2, 1993, pp. 121–130.

major British fighter during the Battle of Britain)." When Mitchell approved the elliptic wing for the Spitfire, he had to have some concern about its manufacturing complexity. Yet he still went ahead with the aesthetically beautiful elliptical wing shape. Shelton simply states: "One suspects that aesthetics and intuition had quite a lot to do with the final choice for the Spitfire wing shape."[21] There is no doubt that Mitchell made changes to engineering drawings by eye and by pure feelings. It is worth repeating here a quote from Joe Smith given earlier:[22] "He was an inveterate drawer on drawings, particularly general arrangements. He would modify the lines of an aircraft with the softest pencil he could find, and then remodify over the top with progressively thicker lines, until one could finally be faced with a new outline of lines about three-sixteenths of an inch thick." Smith went on to state: "But the results were always worthwhile, and the centre of the line was usually accepted when the thing was re-drawn." Mitchell did not make such modifications on the basis of calculations; rather it was pure engineering intuition with an artistic license – a mark of a grand designer.

MORE NEW TECHNOLOGY – THE MEREDITH RADIATOR, WING FILLETS, AND THE MERLIN ENGINE

Much less striking to the eye than the elliptic wing, but perhaps equally if not more aerodynamically important, was Mitchell's use of the Meredith radiator and wing fillets. Both the radiator and the fillets were new technology in the 1930s, but sufficiently proven that Mitchell felt comfortable in using them in the design of the Spitfire.

A young British engineer working for the Royal Aeronautical Establishment in Farnborough, Frederick Meredith, conceived a new radiator concept for use with airplanes powered by liquid cooled engines. The basic cooling mechanism for any liquid-cooled engine (such as the engine in most automobiles) is an initially relatively cool liquid that circulates throughout the engine block, taking heat away from the engine and becoming a hot liquid. This in turn is cooled by passing cool air from the airstream over the hot coils. The cooling air is heated in the process, and ordinarily this hot air simply exhausts out the back of the radiator with an attendant waste of heat energy. Meredith's idea was straightforward. He proposed passing the hot air through a specially designed

[21] Shelton, pp. 212–213. [22] Smith, p. 126.

convergent exhaust duct that acted like a nozzle, increasing the velocity of the exhaust gas, and creating a small forwarding acting force (thrust) on the exhaust duct, thus cancelling some (but not all) of the radiator drag. His theoretical analysis published in an Aeronautical Research Council R & M in 1935 supported the beneficial result.[23] In late 1935, a series of tests of the Meredith radiator were conducted in the Farnborough wind tunnel on a one-fourth scale model, with the radiator located under the standard wing, and the results were "satisfactory." On the basis of the positive theoretical calculations and the wind-tunnel data and after a meeting between Meredith and Supermarine personnel on September 11, 1935, Mitchell chose to use the Meredith radiator on the Spitfire. It was the first airplane to be so equipped. In Figure 5.12d, the radiator cowling can be seen underneath the wing of the final design configuration for the Mark I Spitfire; it does not appear in the earlier drawings in Figures 5.12a, b, and c.

Seen in all four of the drawings in Figure 5.12 is the wing fillet, a smoothly contoured fairing at the wing-fuselage juncture. In the 1930s, the fillet was a new design feature on low-wing airplanes. Without the fillet, the flow over the top of the wing at the fuselage juncture could become separated from the surface, resulting in large fluctuating turbulent eddies that created violent buffeting and increased drag. With the fillet, the flow in the wing-fuselage juncture was much smoother and well-behaved. In the early 1930s, wind-tunnel tests carried out in the large subsonic wind tunnel at Caltech in Pasadena under the direction of Theodore von Karman proved the aerodynamic value of fillets. Indeed, as seen in Chapter 4, the Douglas DC-1, progenitor of the famous DC-3, was designed with fillets, and a model tested in 1932 in the Caltech tunnel. Fillets became a feature of the design revolution in the 1930s, and when Mitchell designed the Spitfire in 1935 with fillets, he was again incorporating relatively new but proven technology.

Perhaps the single most important new technology chosen by both Mitchell and the Air Ministry, however, was the Rolls-Royce Merlin engine for the Spitfire. Following on the heels of Rolls-Royce's powerful but specialized 'R' engine used for the winning Supermarine S.6 and S.6B Schneider Trophy racers, Ernest Hives, chief engineer at Rolls-Royce, committed company resources in 1932 towards the development of a new high-power 12-cylinder liquid-cooled engine for conventional

[23] Meredith, F. W., "Cooling of Aircraft Engines with Special Reference to Ethylene Glycol Radiators Enclosed in Ducts," *ARC, R&M* No. 1683, August 1935.

airplanes. Labeled the PV-12 (PV for private venture), the first drawings were in the shop at Derby in April 1933, and by October the first two engines were ready for tests. At this time the Air Ministry started to provide funding for the new engine, so the PV12 was a private venture for only one year. The first unit passed its type test in July 1934, producing an impressive 790 hp. On November 23, Hives visited Supermarine to discuss with Mitchell and his team the use of the PV12 for the new fighter, labeled by Supermarine as the Type 300. Rolls-Royce and Supermarine had shared a symbiotic relationship since the Schneider racers, and Mitchell and Hives respected each other. The decision to use the new Rolls-Royce engine for the Spitfire was essentially made at that time, and was certainly cemented by the Air Ministry when Specification F37/34 was officially signed on January 3, 1935, allocating 10,000 British pounds to Supermarine for the production of the prototype Spitfire using the PV12 engine. On January 10, 1935, the new engine, now producing 950 hp, was named the Merlin engine by Rolls-Royce. This powerful engine along with the aerodynamic streamlining of the Spitfire, combined to create the fastest fighter airplane in the world at the time. Once again, the Merlin was a case of Mitchell being willing to use new technology in his design of the Spitfire, but here the new technology was only at the ragged edge of being proven. Although designed and built by one of the most reliable aircraft engine companies, the Merlin was still brand new, and the Spitfire was the first production airplane to use it. In spite of this, Mitchell and the Air Ministry together were willing to take the risk.

MITCHELL'S PERSONAL CHALLENGE

At the time he was designing the Type 224, Mitchell was not feeling well physically. In late summer of 1933, he came ill; the doctors told him that he had cancer of the rectum and required an immediate operation. He entered the hospital in August 1933 for a major operation which left him with a permanent colostomy. After convalescing in Bournemouth for a few months, he was back to the design board by December and into the thick of the test preparation for the Type 224. Mitchell, somewhat more reserved now, carried on as usual. Only his close family and his doctors knew about his disability; his colleagues at Supermarine knew nothing about it. Mitchell devoted himself to his work, constantly living with the doctors' information that if he would remain cancer-free for four years, the cancer would most likely not return. Each year, as the anniversary of his operation would come around, Mitchell noted it in his diary. For example, in his August 15, 1936 entry, he simply writes "3 years on." By this time the prototype Spitfire had successfully flown and the Air Ministry had issued Supermarine a contract for the initial production of 310 airplanes.

Unfortunately, through the autumn of 1936 Mitchell began to suffer severe bouts of pain which became more and more frequent. His son wrote:[24] "Day after day Mitchell continued to go in to work, only remaining at home when the discomfort became too severe, and in spite of everything he continued to hope." At this stage, R. J. had moved on to work on a new four-engine bomber design, considering that the Spitfire was now a done deal. By February 1937, however, Mitchell was back in the hospital. After his second operation, all the tests showed there was no hope and he had at best only four or five months to live. He went home and immediately started to arrange his affairs. Again, his son Gordon wrote:[25] "Mitchell wanted to live, but since the Doctor's verdict he was quite prepared to die, and seldom has any man accepted his cruel fate with such coolness and courage. He could no longer go to work, but he received visits from members of his Design Team who wanted to discuss technical problems." As a last effort to overcome his illness, Mitchell visited the world-renowned Cancer Clinic in Vienna, where he received the most advanced treatment. After five weeks of intensive treatment, Mitchell was told that nothing more could be done. Mitchell and his wife Florence returned to England the same way they had come to Vienna, by private airplane.

At noon on June 11, 1937, R. J. Mitchell died at home at the age of 42. His ashes were laid to rest in the South Stoneham cemetery near Eastleigh Airport, where the first flight of the Spitfire had taken place. His death was widely reported in the national press, and tributes were received from throughout the aeronautical world. In Mitchell's obituary that appeared in *The Airplane* on June 16, 1937, the editor C. G. Grey wrote: "In him the British Aircraft Industry has lost one of its most brilliant designers and one of the most valuable and valued members." Among the large number of letters of sympathy received by Florence, the one written by Joe Smith, the person who took over the position as chief designer at Supermarine after Mitchell's death, is perhaps the most descriptive of R. J.'s impact on people:[26]

[24] Mitchell, p. 143. [25] Ibid, p. 144. [26] Ibid, p. 157.

On behalf of the Drawing Office Staff, I have been asked to convey to you and your son our very sincere sympathy in your great bereavement. It was with deep regret that we learned yesterday of this irreparable loss. Mr. Mitchell was much admired and beloved by us all here, and as our leader for a number of years has set us a very high example of courage under trial and of great devotion to duty.

THE OTHER ELLIPTIC WING FIGHTER AIRPLANES

When the K5054 prototype Spitfire with its aesthetically beautiful elliptic wings took off from Eastleigh airport for its first flight on March 5, 1936, it was not the only elliptic wing fighter in the air. In the popular culture of aviation lore, it is the Spitfire that is always recognized by its elliptical-shaped wings. In contrast, two important fighter airplanes designed in the United States by Alexander Kartveli, chief designer for the Seversky Aircraft Corporation – the P-35 (Figure 5.13) and the famous P-47 (Figure 5.14) – both have aesthetically beautiful elliptic wings, a feature that goes virtually unnoticed by aviation buffs, and is rarely pointed out by contemporary writers.

Alexander Kartveli[27] grew up in the Georgian province of Russia and served as an artillery officer in the Russian Army during World War I. As a result of the 1917 Revolution, he fled to Paris where he became interested in powered flight. He was admitted to L'École Supérieure d'Aéronautique where he studied aeronautical engineering. After graduation in 1922, he worked for several airplane manufacturers, the Société Blériot Aéronautique being the most noted. During this period he became familiar with the construction of all-metal airplanes. At the invitation of the American millionaire Charles Levine, Kartveli moved to New York to work on an advanced all-metal monoplane that proved to be underpowered and unsuccessful. After a period of unemployment and poverty, he was hired by the Atlantic Aircraft Corporation, the American subsidiary of the Dutch Fokker Company. In 1931 he met fellow Georgian and by then prominent engineer Alexander de Seversky, who hired Kartveli to work for the fledgling Seversky Aircraft Company on Long Island. By 1933, Kartveli had designed the Seversky SEVC-3, a low-wing amphibian

[27] Libbey, James K., "Alexander Kartvelli: The Aircraft Designer Who 'Suffered' Greatness," *American Aviation Historical Society Journal*, Vol. 56, No. 2, Summer 2011, pp. 131–141.

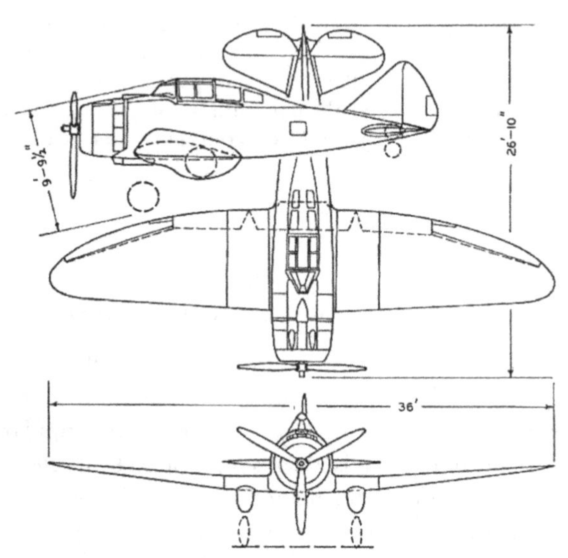

FIGURE 5.13. Three-view of the Seversky P-35, 1935, showing its elliptic wing shape.

monoplane which broke the world speed record for amphibians. Notably, Kartveli had designed the SEV-3 with *elliptic wings*.

The knowledge of Ludwig Prandtl's finite wing theory, which had demonstrated for the first time the aerodynamic efficiency of the elliptic wing planform, quickly spread through continental Europe after World War I, first with French speaking aeronautical engineers at the Belgium Aeronautical Laboratory (now the well-known Von Karman Institute for Fluid Mechanics) just outside Brussels, and then to laboratories in France itself. It is safe to assume that Kartveli was exposed to Prandtl's theory and to the aerodynamic advantages of the elliptic wing during his aeronautical engineering education in Paris. When he designed the elliptic wing for Seversky's SEV-3, there was no other compelling reason than aerodynamic efficiency (and possibly aesthetic reasons, since the SEV-3 was produced for sale to the public). Other than changes in dimensions, this was the same wing that was used two years later on the P-35 (Figure 5.13) and ultimately on the World War II P-47 Thunderbolt (Figure 5.14), by which time the Seversky Aircraft Company had become the Republic Aircraft Corporation.

The Seversky P-35 slightly predated the Spitfire Mark I. The SEV-3, from which the P-35 evolved, first flew in June 1933, almost three years

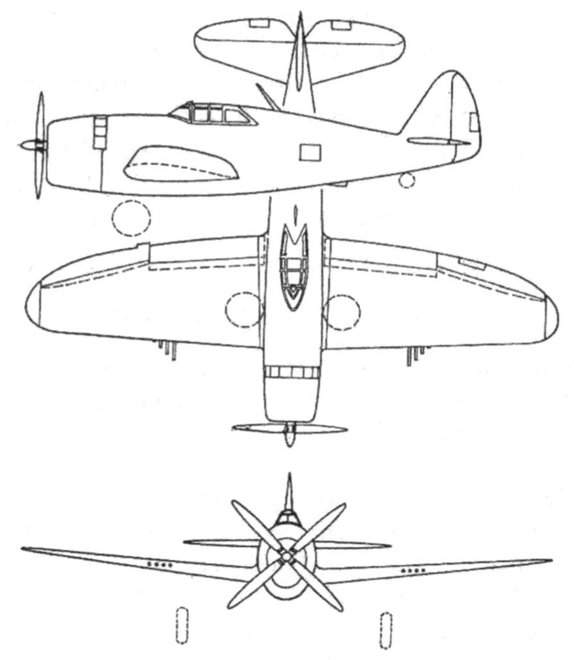

FIGURE 5.14. Three-view of the Republic P-47, 1940.

before the Spitfire prototype K5054. The P-35 arrived at Wright Field on August 15, 1935 under the designation SEV-1XP, commencing a series of fly-off competitions with other airplanes, most notably a Curtiss pursuit which later became the P-36. The fly-offs started seven months before the first flight of the Spitfire. The final fly-off was during April 1936, at which time Seversky was awarded a contract for seventy-seven P35s, about one month after the first flight of the Supermarine K5054. For all practical purposes, therefore, there were two new elliptic-winged fighter airplanes in the air at about the same time, each separated by the Atlantic Ocean. Did the design of one influence the other? Not likely. Kartveli designed his first elliptic wing in 1933, long before the elliptic wing for the Spitfire appeared on paper, and stuck with it through the end of World War II. Had the design team at Supermarine, notably Mitchell and Shenstone, seen a Seversky airplane and been influenced by its wing planform? A glance at Figures 5.12a and b – which show the early configurations in 1934 for the Spitfire with straight tapered non-elliptical wings – dispels the idea of any such influence. The appearance in 1935 of the elliptical wing for the Spitfire, first seen in the drawings of Figure 5.12c, and then in 5.12d, reflects the Air Ministry's edict for an eight-gun fighter, and Supermarine's design approach to make room in the wing for these guns. Nevertheless, Mitchell and Kartveli both had an aesthetic appreciation of the beauty of the elliptic wing and both independently went on to design elliptic-wing fighters that played pivotal roles during World War II: the Supermarine Spitfire, and the Republic P-47 Thunderbolt.

R. J. MITCHELL – POSTSCRIPT

Reginald Joseph Mitchell was a natural "grand designer." He did not pioneer any new intellectual methodology for conceptual airplane design; rather, he quite naturally followed the methodology set forth by Frank Barnwell. But his innate feeling for the creation of successful airplanes, his special intuitive sense of what worked and what did not, his willingness to incorporate new but proven technology in his designs, and simply his creative genius, made him a "grand designer" in the mold of the Wright brothers, Frank Barnwell, and Art Raymond. Different from these other designers, however, Mitchell developed a disdain for wind-tunnel testing in direct support of his design effort. No wind-tunnel data was used for the design of the Spitfire. Mitchell, the "grand designer," did not need it.

As a final note in regard to the conceptual design of the Spitfire, it included the use of the powerful Rolls-Royce Merlin engine, the elliptical wing with the wing fillet, the NACA airfoil shape, and the Meredith radiator well before the Spitfire design moved to the next step, that of preliminary design where only fine-tuning takes place. The inclusion of new technology was an integral part of Mitchel's conceptual design, but still using Frank Barnwell's original methodology. Here, the conceptual design methodology and the immediate use of new technology were integral with each other. Moreover, Mitchell's use of the new technology was not completely a matter of his intuition; rather, he was a Fellow of the Royal Aeronautical Society and was in contact with aeronautical engineers at the Royal Aeronautical Establishment. Through these contacts, as well as his voracious reading of the technical literature, he was well aware of what was new in aeronautical engineering.

Among the many tributes to R. J. Mitchell that followed long after his death was this one, perhaps the best, from Group Captain David Green, chairman and founder of the postwar Spitfire Society who wrote:[28]

[28] Gordon Mitchell, p. 158.

Throughout my flying years my curiosity has grown consistently about the man whose brain spawned what has always been to me an artistic masterpiece even before its engineering excellence. The latter was always present in the work of R. J. The Southampton, the Stranraer, the Walrus were all fine examples of his engineering skill. But the explosion of aesthetic genius, first with the S.4 and culminating with the Spitfire, was the ingredient which raised the man to a level unattainable by the most brilliant of his contemporaries.

6

Design Perfection

Edgar Schmued and the P-51 Mustang

The North American P-51 Mustang is considered by many to represent the highest level of technical refinement ever achieved in a propeller-driven fighter aircraft.

Laurence K. Loftin, Jr., *Quest for Performance: The Evolution of Modern Aircraft*

The scene: the North American ramp at Mines Field, Inglewood, California. *The time*: the morning of October 26, 1940. *The characters*: the design team for North American Aviation (NAA), including the chief designer, Edger Schmued, a number of plant workers, and Vance Breese, an NAA test pilot. *The action*: The group is huddled around a strikingly beautiful new fighter airplane, its unpainted silvery-looking aluminum surface gleaming through the California autumn haze. The airplane is the North American NA-73X, a hand-built prototype of the company's first fighter design (Figure 6.1). Designed by Edgar Schmued, the airplane is exceptionally clean aerodynamically, and exhibits some modern, even relatively unproven technology. The airplane is powered by an Allison V-1710–39 in-line liquid-cooled engine. Like the Merlin engine that powers the Spitfire, the Allison engine provides the most power with the least frontal area, and allowed Schmued the luxury of designing a highly streamlined shape for the NA-73X. A chase plane takes off in anticipation of making close observations of the NA-73's first flight. Last minute consultations take place on the ground. Then Breese starts the engine and taxis to the end of the runway, pausing there to run the engine through a power check. It had rained during the night, and the NA-73 kicks up a cloud of mist. As Breese releases the brake and gradually

FIGURE 6.1. Prototype North American XP-51 (NA-73X) fighter.
(National Air and Space Museum, SI-2003-4964.)

increases the throttle, the airplane rolls down the runway. The 1,120 horsepower Allison engine roars smoothly as the airplane lifts into the air. The airplane that is destined to become one of America's most famous World War II fighters – the P-51 Mustang – is flying for the first time. The ninety-seven engineers and technicians on the ground cheer and congratulate themselves, especially because the airplane flying above them had been designed and built completely with company funds. Breese tests the airplane for about twenty minutes, putting it through a series of maneuvers – all within sight of the airport. The airplane's performance is excellent, exceeding all the preliminary calculations. After falling into formation with the chase plane, Breese brings the NA-73 in for a smooth landing and taxis to the same ramp from which it departed. Like the completion of the Spitfire's first flight, there was hardly anything that needs to be touched. The success of the prototype NA-73 is a graphic testimonial to the genius of its designer, Edgar Schmued. The stunning success and impact of the more than 5,000 P-51 Mustangs that follow launches Schmued into the pantheon of the "grand designers." Just how this happened is the story told in this chapter.

AIRPLANE DESIGN – STATE OF THE ART IN 1940

The impact of Edward Warner's seminal book *Airplane Design: Aerodynamics*, 1927 on the intellectual process of airplane design during the design revolution of the 1930s is discussed in Chapter 4. In 1936 Warner published a second edition entitled *Airplane Design: Performance*.[1] Like the first edition, this book is a presentation of the updated existing technology available for airplane design as well as a reflection of the contemporary intellectual approach made by airplane designers on the use of this new technology.

An important case in point is the use of wind-tunnel data by airplane designers. The Wright brothers' wind tunnel provided insight about the beneficial effect of increasing the wing aspect ratio, resulting in the doubling of the aspect ratio for their 1902 glider compared to their earlier designs, and leading to success. Frank Barnwell used airfoil data obtained from the National Physical Laboratory for his Bristol fighter and campaigned successfully for the British Airplane Company to build their own wind tunnel in 1919. In 1932 tests of a model of the Douglas DC-1 in the Caltech wind tunnel showed the airplane to be longitudinally unstable, resulting in the redesign of the wing to incorporate a mild degree of sweepback to reposition the center of lift. In spite of these rather pointed examples, however, airplane designers were leery of using wind tunnel data for their conceptual designs, and for good reason. The early wind tunnels for the most part did not produce accurate, reliable data. Also, with the exception of the DC-1 model in the Caltech wind tunnel, no wind tunnel models of a complete airplane configuration were used by the Wrights, Barnwell, or Mitchell to assist in their conceptual design process. The scale effects and lack of quality air flow in the early wind tunnels could have tainted any such data if it had been obtained. The grand designers during this period simply forged their creations as if wind tunnels did not exist.

Warner's second edition in 1936 goes a long way toward changing this negative attitude about wind tunnels on the part of airplane designers. In 1936 there existed several relatively new wind tunnels that greatly negated the scale effect problem. In particular, at the NACA Langley Memorial Aeronautical Laboratory, the Full-Scale Tunnel became operational in May 1931.[2] This tunnel had a test section measuring 30 feet by 60 feet in cross section, allowing whole airplanes to be mounted in the tunnel. Nine years before the Full-Scale Tunnel, in October 1922, the Variable Density Tunnel[3]

[1] Warner, Edward P., *Airplane Design: Performance*, 1936, McGraw-Hill Book Co., New York

[2] Hansen, James R., *Engineer in Charge: A History of the Langley Aeronautical Laboratory*, 1917–1958.

[3] Ibid, p. 443.

came online at Langley, pressurized to twenty atmospheres and allowing a twentyfold increase in Reynolds number. These facilities, as well as others built in Europe, went a long way toward providing more reliable data for airplane designers. Warner emphasized this point in his new second edition. Reflecting on the new wind tunnels, he says:[4]

> All this refinement becomes increasingly more important, and engineers become increasingly exigent in their demands for it, for two reasons. First, because as design becomes increasingly an exact science, and as designers are required to subject themselves to ever closer specifications and advance guarantees for their product's performance, errors of 5 or 10 percent in aerodynamic data, that might once have seemed too small to notice, come to assume nightmare proportions.

The second reason he gives is:

> Because the accumulation of experience has shattered such comfortable illusions, once commonly held, as that the variation of aerodynamic coefficients with Reynolds number is in practically every case a smooth curve, that in such variations as exist there is an asymptotic approach to a constant level at high Reynolds numbers, that sudden and violent changes of behavior in any case do not occur there, and that the data obtained at a Reynolds number of one-tenth full scale or thereabouts can therefore be taken as valid for any higher figure. Those things unfortunately are not true.

About the scale effect, Warner states:

> The more completely it can be eliminated by using test data taken at the Reynolds number at which the aircraft is finally to be operated the better off the designing engineer will find himself.

The wind tunnels available in the 1930s went a long way towards closing this gap.

Note that Warner now calls airplane design "increasingly an exact science." This was another aspect of the design environment of the 1930s in comparison to the previous decades. Airplane designers were becoming more aware of the value of engineering science, especially that of contemporary aerodynamic theory, in forming their design thinking. In his second edition in 1936, Warner reflects on the environment that existed nine years earlier when his first edition was published. "In 1927, except for an occasional and gradually increasing use of induced-drag formulas, the typical designer thought of himself as having very little use for the offerings of the mathematicians, and he had very little disposition even to

[4] Warner, p. 27.

inquire into their application to this own work. He is much more disposed in that direction now."[5]

Perhaps the best indication of the state of the art of airplane design in the late 1930s is that the first flight of the prototype Spitfire in 1936 was extremely successful, as was the first flight of the prototype Mustang in 1940. Is this because both Mitchell and Schmued were gifted airplane designers who were experienced and knew what they were doing? Yes. Is this because these men incorporated new but reasonably proven technology in their designs? Yes. Is this because they used a completely new intellectual methodology in forming their conceptual designs? No. Indeed, both Mitchell and Schmued followed the same design methodology laid down by Barnwell in 1916. In contrast to the design environment during World War I, when airplane designers breathed a sigh of relief when their creations simply got off the ground, flew, and landed safely, in the late 1930s properly designed airplanes not only got off the ground for their first flight, but performed maneuvers and flew at reasonable speeds before landing. This was the state of the art of airplane design in the late 1930s.

EDGAR SCHMUED – THE MAN

The P-51 Mustang was designed by an immigrant from Germany. Edgar Schmued was born on December 30, 1899, in Hornbach, a small town of about 1,200 people near Zweibrucken and the border with Lorraine. His father, a dentist desirous of a larger clientele, moved his family to Landsberg on the Warthe, a community of 35,000 east of Berlin. One of six children in the family, Edgar grew up in Landsberg, attending middle school and then business school. "I was not a good student," Schmued wrote much later, "because my interests were not in homework and schoolwork, but in technical developments."[6] Because of his poor record in school and because his father did not have the financial resources, Schmued did not attend college. Instead, he was self-taught, encouraged by his father to read extensively in technical books from the library. As a boy he was especially interested in aviation, triggered by the pre-World War I success of the Wright brothers and others. He was 12 years old when just by chance while on a family outing, he saw an airplane in flight for the first time. It was the Russian aviator, Vsevoled Abramovich, in a

[5] Ibid, p. vi.

[6] Schmued, Edgar, unpublished manuscript, 1985. (See Wagner, Ray, *Mustang Designer*, Orion Books, New York, 1990, p. 26).

homebuilt Wright-type airplane on his way from Berlin to St. Petersburg. "It made such a tremendous impression on me," Schmued wrote, "that I decided right then and there, that this was for me. This was going to be my life."[7]

Schmued was a self-educated person. In lieu of not going to a university, he mentally devoured technical books, studying to learn engineering. "Although it was difficult" Schmued reflected many years later, "very difficult, it strengthened my character and I succeeded in gathering enough knowledge to step out and eventually become an airplane designer."[8]

Edgar became an apprentice in a small engine factory when his schooling ended. There he learned machining, working on a large lathe, milling machines, and shapers. By the end of his two-year apprenticeship, he had finished a project given to him by the factory owner, the building of an engine completely from scratch. By March 1917, Edgar found himself in the Austro-Hungarian Flying Service as a mechanic. His service was brief, however, due to the surrender on November 3, 1918. From there, Schmued returned to Landsberg, where his father helped to provide resources for the design of a small, homebuilt sports biplane; construction was initiated in his father's library. Because of the severe restrictions put on the German construction of aircraft by the Versailles Treaty, the Allied Control Commission took away the small engine that Edgar had planned to use, and that ended his attempt to design and build his first airplane.

In 1921 Edgar went to work for a developer of automobile equipment in Bergedorf, a suburb of Hamburg. Within a year, he had obtained five German patents dealing with various fuel devices. By this time he was married, and he and his wife Luisa had a son, Rolf Dietrich Schmued, born on June 14, 1921. During the 1920s the Weimar Republic in Germany suffered from massive inflation and political instability. Edgar's oldest brother, Erwin, had followed his father into dentistry and had immigrated to São Paulo, Brazil, where a fairly large German community had formed. The youngest brother Erich, also a dentist, was next to move to São Paulo, soon to be followed by Edgar in 1925 with visions of building on his interest in airplanes. He found, however, no one in Brazil interested in starting a venture into the airplane business and wound up working for a General Motors automobile agency as a mechanic. But not for long. Impressed by Edgar's invention of ways to improve the service of automobiles, a General Motors executive moved him to the main Brazil

[7] Ibid, p. 27. [8] Ibid, p. 28.

office and put him in charge of the organization for field service and agencies. Serendipity took over. In 1929 General Motors became the controlling interest in the Fokker Aircraft Corporation of America, and in February 1930 Edgar immigrated to America to take a position with Fokker, located at the Teterboro Airport in New Jersey. In spite of the 1929 stock market crash Fokker was doing well, producing the three engine F-10 transport and the four engine F-32 transport. Prototypes for observation and attack planes for the US Army were in the works. Unlike most American airplane companies which were still building mainly biplanes, Fokker was in the forefront with monoplanes. Schmued found himself in a good place.

At Fokker, Edgar was able to blossom as an airplane designer. "I felt that companies in America didn't have the proper organization by not having preliminary design departments," he said later, "so I started the first preliminary design department in the US as part of Fokker Aircraft, and as such, I did a great deal of design work for new models."[9] Schmued, of course, was too new to be the chief designer, the position held by Anthony Fokker. But corporate changes were on the horizon. On May 24, 1930 General Motors formed the General Aviation Corporation of America, with Anthony Fokker continuing as chief engineer.

On March 31, 1931 disaster struck the Fokker company when a TWA Fokker F.10 trimotor crashed in a Kansas wheat field, killing among the passengers the famous Notre Dame University football coach Knute Rockne. Fokker airplanes were of wood construction, and the TWA airplane's main wooden wing spar had delaminated, causing wing failure. This highly publicized event not only hastened the end of the wooden airplane, it suddenly brought new Fokker transport sales to an end. Although the company was designing new all-metal airplanes at the time, it could not survive. Fokker resigned and the New Jersey plant was closed. What was left of the General Aviation operations moved to Dundalk, Maryland, near Baltimore. Schmued and his family moved with it.

The enterprise at Dundalk floundered. It was kept afloat initially by a few small airplane orders for a Navy fighter, the XFA-1, and the Army observation plane, the OX-27, a twin engine monoplane with retractable landing gear. Schmued was made the project engineer for the OX-27, but in the end the Army bought only twelve of these aircraft. Then he was made the project engineer for the GA-43, a rather ungainly-looking

[9] Ibid, p. 31.

all-metal, low wing, single-engine monoplane initially designed by Virginius Clark, an illustrious figure in the 1920s and 1930s in aviation. A colonel in the US Army, Clark at one time commanded McCook Field, the Army's aeronautical testing facility in Dayton, Ohio. Clark is perhaps best known for his design of a successful series of Clark airfoils, the Clark Y, in particular finding use on a large variety of US and foreign aircraft. He was not so successful, however, as an airplane designer. Schmued wrote about his experience with the GA-43: "Although it was not a very striking piece of design, it gave me at least a chance to organize and run a project."[10] To make things worse, by this time the airlines were no longer interested in single-engined airplanes and the carrying capacity of the GA-43 was only ten passengers. Manufacture of the GA-43 began in 1934; only five were built.

In 1934, however, something much more important happened to General Aviation. James Howard (Dutch) Kindelberger, formerly the vice president and chief engineer of Douglas (see Chapter 4), became president of General Aviation on July 13. Six months later the company took the name of the stockholding company that controlled its stock – North American Aviation. Under the Air Mail Act of 1934, this stockholding company was required to become an operating company. Thus, the airplane manufacturing firm, North American Aviation, came into being. In this sense, the ill-fated GA-43 became the first airplane to be produced by North American, and the company's first project engineer was Edgar Schmued, almost by default.

Because the center of activity in the aircraft industry was rapidly focusing on Southern California, and because the weather there was much more favorable to flying than in Maryland, Kindelberger made the decision to move North American to Los Angeles. In September 1935 the company opened a shop on a 20-acre site beside Mines Field, later to become the Los Angeles Municipal Airport. On November 1 of that year, with new orders in hand from the Army for a new basic trainer, the BT-9, North American broke ground for a large factory at the corner of Imperial Highway and Aviation Boulevard in Inglewood, a suburb of Los Angeles.

By the end of 1935, approximately seventy-five workers had relocated from Dundalk to Los Angeles, and another 175 new workers were hired by North American. Edgar Schmued, however, was not among them. His wife, Luisa, wanted to stay on the east coast (it was closer to her native

[10] Ibid, p. 33.

Germany than California). For a while, Edgar worked at Bellanca Aircraft in New Castle, Delaware. In the process, he became an American citizen on October 21, 1935. It did not take him long to realize that Bellanca was not on the forefront of aeronautical technology and offered very little promise for an ardent designer of airplanes. With Luisa's realization that California and North American was the best opportunity for Edgar, the family packed up and drove cross-country.

Disaster struck! They were barely across the California state line on Route 60 when they were struck head-on by another car. Luisa was killed; Edgar and his son were badly injured. It was November 12, 1935. Edgar was in the hospital for two months, recovering from a concussion, a massive fracture of his leg, and facial injuries. It was not until February 1936 that he was able to start work again with North American.

Schmued was one of those people who could bury his feelings under a blanket of hard work. North American had just begun the design of a new twin-engine bomber, the NA-21. Edgar was put to work designing the hydraulic power-driven turret for the airplane; it was the first such turret on a US Army Air Corps plane.

The NA-21 proved to be too slow, and the Air Corps ordered instead the Douglas B-18, a bomber version of the DC-3 (see Chapter 4). Nevertheless, in the next four years leading up to the United States' entry into World War II, North American Aviation grew and prospered, mainly on the shoulders of some very successful trainers, from the BT-9 in 1936 through the AT-6 Texan in 1940. The AT-6 (Figure 6.2) was destined to become the best known training aircraft of the War; many AT-6s are still flying today as sport, aerobatic, and purely collector's airplanes. And even though the NA-21 was not successful, North American's follow-on design of a twin-engine bomber in 1939, the NA-62, became the famous B-25 Mitchell (Figure 6.3) of which 10,000 were produced before the end of the War.

Ed Schmued participated in the preliminary design of the B-25. By 1940 he was chief designer for North American and poised to take on the design of a new fighter that would elevate his status to a "grand designer." That new fighter was the P-51 Mustang.

GENESIS OF THE P-51 MUSTANG

In early 1938, the British government realized that their domestic production of fighter airplanes was not going to satisfy the need for the total defense of Britain against German air attacks. The British had an excellent

FIGURE 6.2. North American AT-6 Texan.
(National Air and Space Museum, NASM-7A35039.)

FIGURE 6.3. North American B-25 Mitchell.
(National Air and Space Museum, SI-91-2424.)

fighter in production at the time, the Spitfire (see Chapter 5), as well as the slightly older and slower Hawker Hurricane. But in war, sheer numbers of fighting machines was frequently as important as the quality of performance, and Britain did not have the numbers. In March of 1938, Britain organized a purchasing commission under the direction of A. T. Harris and Sir Henry Self to travel abroad and buy, among other weapons, warplanes from other countries. On November 7, the British established a New York office of the purchasing commission under the direction of Arthur B. Purvis. The US Congress had just reversed earlier neutrality embargoes and Purvis proceeded to order airplanes from a variety of American manufacturers. Included was an order of 400 AT-6 trainers from North American, later to be named the "Harvard" by the British.

FIGURE 6.4. Curtiss P-40C.
(National Air and Space Museum, NASM-2A15169.)

FIGURE 6.5. Bell P-39.
(National Air and Space Museum, SI-80-15298.)

At this time the only American fighters in production acceptable to the British were the Curtiss P-40 Warhawk (Figure 6.4) and the Bell P-39 Airacobra (Figure 6.5). The Supermarine Spitfire was a better fighter than either of these two American designs but the British needed numbers. These two airplanes had already been ordered by the Armée de l'air of France and were also in production for the US Army Air Corps. When the commission placed an order for 560 Warhawk aircraft from Curtiss, the company's production line was already saturated and Curtiss could not start working on the British contract until May 1, 1940.

This was a problem for the British. Sir Henry Self knew Dutch Kindelberger personally, and on February 25, 1940 he asked Kindelberger if

North American could help by building additional P-40s. This did not sit well with the president of a now thriving aircraft manufacturer. He went to his chief designer. It turned out that Schmued was already thinking about a new fighter airplane and had, on his own, already prepared a preliminary design for a future fighter, which he had promptly stashed away in his desk drawer. In Edgar Schmued's own words:

> One afternoon in March, Dutch Kindelberger came to my office and said, 'Ed, do we want to build P-40s here?' From the tone of his voice, I knew what kind of an answer he expected. I said 'Well, Dutch, don't let us build an obsolete airplane, let's build a new one.' And that's exactly what he wanted to hear.

Of course, Schmued was speaking with the knowledge that the first conceptual idea for a "better one" was on paper, sitting in his desk drawer. Schmued continued:

> So he said, 'Ed, I'm, going to England in about two weeks and I need an inboard profile, three-view drawings, performance estimate, weight estimate, specifications, and some detail drawings on the gun installation to take along. Then I would like to sell that new model airplane that you develop.' That was the order I got from Dutch Kindelberger.

Kindelberger was an experienced aeronautical engineer and what he had just asked of Schmued was a conceptual airplane design using the tried and proven intellectual method as prescribed by Frank Barnwell in 1916 (see Chapter 3). Schmued continued:

> He said that the rules for design were simple. Make it the fastest airplane you can and build it around a man that is 5 feet 10 inches tall and weighs 140 pounds. It should have two 20-mm cannons in each wing and should meet all design requirements of the (Army).

Through the efforts of Kindelberger and his vice president Leland Atwood, and based on Schmued's conceptual design of a new fighter, the British dropped their insistence that North American build P-40s, and on April 11, Sir Henry Self signed a letter of intent to purchase the North American fighter. On April 24, a group of North American engineers under the direction of Ed Schmued met to organize the detailed design and fabrication of their new fighter, now designated the NA-73. On May 4, the British Air Purchasing Commission officially approved the preliminary design, and on May 23 a formal contract was signed between the Commission and North American Aircraft for 400 Model NA-73s. At that time Schmued and North American swung into action seven days a week. Five months later the prototype NA-73 flew for the first time. In regard to the subsequent flight testing of the NA-73, Schmued reflected that "Relatively little was changed, of minor importance, because the plane performed so well." Four and a half years earlier, Spitfire test pilot Mutt Summers, in the same spirit of elation, said at the completion of the first flight of the Spitfire, "I don't want anything touched." (See Chapter 5.) Devoid of any national insignia, and with only its Civil Registry number NX19998 painted on its wing, but for all practical purposes paid for by the British, the prototype NA-73 was the first of a long line of P-51 Mustangs that, like the Supermarine Spitfire, was destined to become one of the most famous and iconic fighter airplanes of World War II. And like the Spitfire, its success was largely due to its designer.

DESIGN OF THE MUSTANG – EMBODIMENT OF NEW TECHNOLOGY

For the conceptual design of the Mustang, Ed Schmued followed the same general intellectual methodology as outlined by Frank Barnwell in 1916. The North American aerodynamicist Edward Horkey wrote in 1996:[11]

> The standard procedure in those days was to have Ed Schmued and his group make an inboard profile and three-view drawing and the start of a specification. Harold Raynor would come in and do the weight work and Aerodynamics would do the performance estimates.

There was nothing new here. What was new, however, was the modern aeronautical technology that Schmued adventurously incorporated in the P-51 during the *detailed* design process, well after the conceptual design process had been completed. This new technology is detailed in the following sections.

The Laminar Flow Airfoil

The selection of a shape of an airfoil section has always been an essential item in the intellectual methodology of conceptual airplane design, starting with the Wright brothers. The primary motivation behind the Wrights' wind tunnel testing during 1901 and 1902 was to find the most effective airfoil shape for their next glider design. Orville Wright, in 1920, wrote:

[11] Horkey, Edward, "The P-51, The Real Story," *American Aviation Historical Society Journal*, Vol. 41, No.3, Fall 1996, p. 179.

In order to try to satisfy our own minds as to whether the failure of the 1900 machine to lift according to our calculations was due to the shape of the wings or to an error in the Lilienthal tables, we undertook a number of experiments to determine the comparative lifting qualities of planes as compared with curved surfaces and the relative value of curved surfaces having different depths of curvature.[12]

Clearly, airfoil shape became an important aspect of their subsequent designs (see Chapter 2). In Frank Barnwell's book in 1916 which established the "gold standard" for the intellectual methodology of conceptual airplane design (see Chapter 3), more than twelve pages are devoted to airfoils and airfoil selection.[13] Barnwell notes that: "Data for aerofoils are founded entirely upon experimental work." He goes on to stipulate that the choice of an airfoil shape, with knowledge of its aerodynamic characteristics must be made before commencing the "actual design" of the airplane.[14] Warner, in 1936, in his second edition of *Airplane Design*, had four chapters on airfoils, wherein he stated: "Whatever the means adopted for representing the characteristics of (airfoil) sections, some essential features of performance must be singled out to be used as a basis of a comparison by the designer choosing a section for a new machine."[15] The importance of airfoil selection in the conceptual airplane design methodology continues to the present day. In his book *Aircraft Design: A Conceptual Approach*, which today is the standard text for most college level airplane design courses, Dan Raymer writes: "Before the design layout can be started, values for a number of parameters must be chosen. These include the airfoil ... The airfoil, in many respects, is the heart of the airplane."[16]

When Edgar Schmued put pencil to paper at the start of the conceptual design of the P-51 in early 1940, the state of the art in airfoil shapes resided at the NACA Langley Memorial Aeronautical Laboratory in Hampton, Virginia. Through the 1920s, the design of airfoils in the United States and elsewhere was essentially ad hoc, leading to a great proliferation of different airfoil shapes but only fragmentary understanding of

[12] McFarland, Marvin W., Ed., *The Papers of Wilbur and Orville Wright*, Vol. 1, McGraw-Hill, New York, 1953, p. 547.

[13] Barnwell, F. S., *Aeroplane Design*, New York, Robert M. McBride and Co., 1917, pp. 21–33.

[14] Ibid, p. 21. [15] Warner, p. 63.

[16] Raymer, David P., *Aircraft Design: A Conceptual Approach*, 4th ed., American Institute of Aeronautics and Astronautics, 2006, p. 37.

basic airfoil aerodynamics.[17] That situation changed dramatically in the early 1930s. Under the direction of Eastman Jacobs, a young engineer who joined the NACA Langley Memorial Aeronautical Laboratory in 1925, one year after graduating with honor from the University of California at Berkeley, the NACA began a systematic program of designing and testing a family of aerodynamically efficient airfoil shapes labeled the "four-digit series." The airfoil shape that Art Raymond used on the DC-3 (see Chapter 4) was one of the four-digit series, namely the NACA 2215 at the wing root, tapering to the NACA 2206 at the tip. (In the four-digit labeling system, the first digit gives the maximum camber in hundredths of the chord length, the second digit gives the location of the maximum camber in tenths of the chord length measured from the leading edge, and the last two digits give the maximum thickness of the airfoil in hundredths of the chord length. For example, for the NACA 2215 airfoil, the maximum camber is 0.02c, the location of the maximum camber is 0.2c behind the leading edge, and the maximum thickness is 0.15c, where c denotes the chord length measured from the leading edge to the trailing edge of the airfoil). For the Spitfire, R. J. Mitchell used an NACA 2412 airfoil at the wing root, tapering to an NACA 2205 airfoil at the wing tip (see Chapter 5). Interestingly enough, the adversary of both the Spitfire and later the P-51 Mustang, the German Messerschmidt 109, incorporated a rather obscure NACA airfoil, the 2R1 14.2 at the root tapering to a 2R1 11.35 at the tip. Another adversary, the German FW-190, also was designed with an NACA airfoil, one of the later "five-digit series," the NACA 23015 at the root tapering to the NACA 23009 at the tip. NACA airfoil shapes became the "gold standard" in the 1930s, and were widely used for airplane designs in a number of countries. Edgar Schmued was no exception; the airfoil shape originally chosen for the P-51 during the conceptual design was also a standard NACA five-digit airfoil.

But a very special "high technology" airfoil ultimately was used for the P-51, a laminar flow airfoil, developed by the NACA. This airfoil was of such importance that it was initially classified secret. The shapes of these airfoils were designed to encourage the boundary layer, a very thin region of the airflow adjacent to the surface where the influence of friction was dominant, to remain a nice smooth laminar flow over a large portion of the airfoil before transiting to a turbulent boundary layer. The skin friction produced by a laminar boundary layer is much smaller than that

[17] Anderson, John D., Jr., *A History of Aerodynamics*, Cambridge University Press, New York, 1997, pp. 342–352.

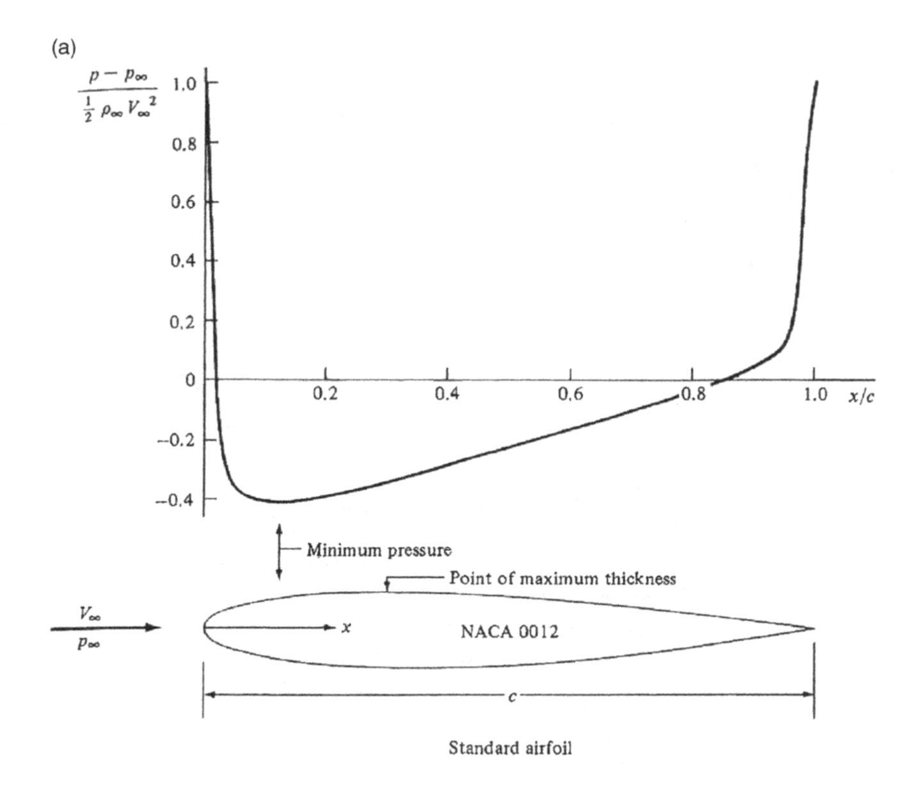

FIGURE 6.6. (a) NACA 0012 airfoil shape and pressure distribution. (b) NACA 66-012 airfoil shape and pressure distribution.

produced by a turbulent boundary layer, and hence an airfoil with an extensive region of laminar boundary layer flow will have a much smaller friction drag than an airfoil swathed by a turbulent boundary layer. To encourage a large region of laminar flow, the NACA designed a family of airfoil shapes to have maximum thickness much further downstream of the leading edge than conventional airfoils. A comparison of a laminar flow airfoil shape, that for an NACA 66-012 airfoil, with that for a conventional four-digit NACA 0012 airfoil, is shown in Figure 6.6, along with the measured surface pressure distributions for both airfoils. Note that there is a much longer distance over the NACA 66-012 airfoil where the pressure is decreasing (called a "favorable" pressure gradient) than over the NACA 0012 airfoil. Favorable pressure gradients encourage laminar flow; hence the boundary layer over the NACA 66-012 airfoil

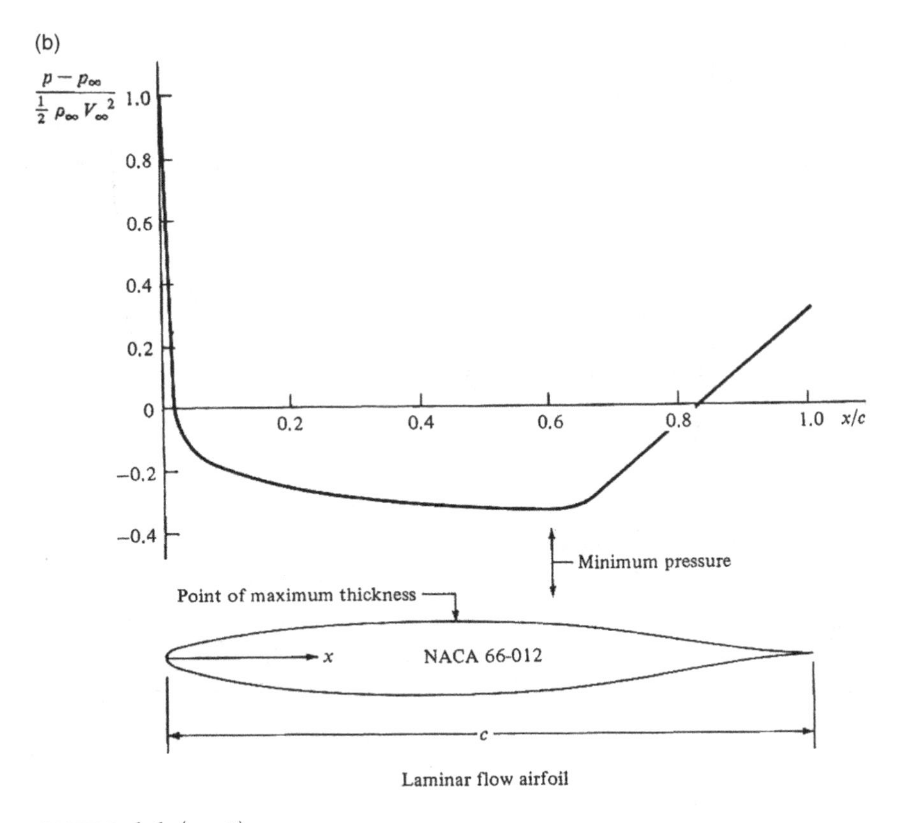

FIGURE 6.6. *(cont.)*

remains laminar at least to the point of minimum pressure, which occurs at about 0.6 of the chord length. By comparison, the NACA 0012 airfoil has its minimum pressure point at about 0.15 of the chord length, with a corresponding smaller region of favorable pressure gradient, hence a much smaller region of laminar boundary layer, and therefore higher friction drag. Spearheaded by Eastman Jacobs at the NACA Langley Memorial Laboratory, a family of laminar flow airfoil shapes was developed and tested in 1938 and 1939. Security restrictions prevented the NACA from going public with Jacobs' new findings; indeed, the actual data on laminar flow airfoils were not released until after World War II.

Edgar Schmued and his aerodynamicist Ed Horkey were informed about the NACA laminar flow airfoil work when Russ Robinson, an

NACA aerodynamicist, and Ed Hartman, the local engineering representative and coordinator for the NACA, visited North American in early 1940. The early plans were to use an NACA 230 series five-digit airfoil on the new North American fighter. The conceptual design process was finished. However, during the detailed design of the P-51, Ed Horkey was particularly intrigued by the concept of the laminar flow airfoil. Rather than use explicitly one of the NACA-designated laminar flow sections, Horkey and his aerodynamics team modified the shape slightly, designing a thinner airfoil to better match their wing design requirements.

Schmued and Horkey, in spite of being venturesome with the laminar flow airfoil, still hedged their bets. The P-51 would be, at that time, the only airplane to have incorporated a laminar flow airfoil; there was no base of experience other than the laboratory tests made on small models by the NACA. Thus, North American began a series of wind-tunnel tests that opened a new vista in the use of wind-tunnel testing for conceptual airplane design. Two scale models of the P-51 wing design were built with the same wing area and planform, but one with the originally planned NACA 230 five-digit airfoil, and the other with North American's modified laminar flow airfoil, called the NACA/NAA 45-100 airfoil section. On May 21 and 22, 1940, the two wing models were tested in the 10-foot wind tunnel at the Guggenheim Aeronautical Laboratory at Caltech (GALCIT). The wing with the laminar flow airfoil produced less drag. However, Schmued was not convinced by the results, concerned that the wing model was too large for the test section and that turbulence along the tunnel walls compromised the aerodynamic characteristics at the wing tips. The laminar flow wing model was then rushed to the larger 8 feet by 12 feet wind tunnel at the University of Washington in Seattle. Horkey later related:[18,19]

> I got on a DC-3 and went up to Seattle with the model tucked away in the back of the airplane. We tested for both [sic] maximum lift, drag, and stall characteristics. And they came out great, so that was our shape from then on.

The P-51 Mustang wing remained the same throughout the many variants of the airplane produced during the War.

[18] See for example, Anderson, John D., Jr., *Introduction to Flight*, 7th ed., McGraw-Hill, New York, 202, pp. 228–251.

[19] Wagner, Ray, *Mustang Designer*, Orion Books, New York, 1990, p. 56.

Edgar Schmued's belief in and reliance on these wind-tunnel tests set a new standard for future airplane designers. No longer was wind-tunnel data looked upon as unreliable. The wind tunnels that went into operation during the middle and late 1930s were high technology themselves, and they contributed greatly to the modern airplane designers putting their arms around new (but proven) technology in their designs. It was this attitude that contributes to Edgar Schmued being in the ranks of the grand designers.

This was not the only wind-tunnel testing that factored into the design of the P-51. From June 28 to July 26, North American returned to the Caltech wind tunnel with a beautiful one-quarter scale, highly polished, mahogany model of the NA-73 prototype. Dr. Clark Millikan, professor of aeronautics at Caltech, had developed wall corrections for the wind-tunnel data. This new set of data further verified the effectiveness of the airfoil shape, and also indicated low drag of the radiator.

The Meredith Radiator (Again)

The Meredith radiator was new technology when R. J. Mitchell decided to use it for the Spitfire (see Chapter 5). It was still new technology when Edgar Schmued chose it for the Mustang. F. W. Meredith was an engineer working for the Royal Aeronautical Establishment in England. In 1935 he published a report mathematically proving that airflow into a ducted radiator, when heated, could result in a jet of air out the duct exit that provides a component of thrust, partly negating the usual aerodynamic cooling drag. The Spitfire was designed with two such radiators, each one mounted on the bottom surface of each wing. Schmued and his team chose to place a single radiator duct integral with the bottom of the fuselage with the inlet almost at the trailing edge of the wing (see Figure 6.1) in an effort to minimize aerodynamic interference with the wing. The above-mentioned quarter-scale wind-tunnel tests at Caltech verified the low radiator drag.

Streamlined Design

As reflected in the P51D shown in Figure 6.7, the aesthetic beauty of Schmued's design matched that of the Supermarine Spitfire and the Douglas D-3 discussed in previous chapters. Although the concept of streamlining had been utilized for the past decade, Schmued took it to new

FIGURE 6.7. North American P-51D Mustang.
(National Air and Space Museum, SI-83-8782.)

heights by using mathematical conic sections to design all parts of the airplane except the wing and tail. As Schmued told it:[20]

> This is the kind of shape the air likes to touch. The drag is at a minimum and it was the first time that a complete airplane, with the exception of the lifting surfaces, was designed with second-degree curves. I laid out the lines myself, and it was a first.

All told, the laminar flow airfoil, the Meredith radiator, and the attention to streamlining resulted in the P-51D having a zero-lift drag coefficient of 0.0163, the lowest of all the World War II fighter airplanes.

The Spitfire and the Mustang shared more than their aesthetically beautiful design, and more than the use of the Meredith radiator, they also shared the expertise of Supermarine's aerodynamicist, Beverly Shenstone. Schmued had designed the radiator so that the upper lip of the inlet was flush with the bottom contour of the fuselage, as seen in Figure 6.1. As a result, the boundary layer that entered at the top of the inlet originated at the nose of the airplane, and had a long running length along the bottom of the fuselage before entering the top inlet lip of the radiator. This portion of the boundary layer was "tired," robbed of much of its energy by friction acting all along its long path from the nose of the fuselage to the radiator itself. This distorted the pressure distribution inside the radiator because the "tired" boundary layer experienced a loss

[20] Ibid, p. 57.

of total pressure, which is a measure of the capability of the air flow to do useful work. There was a rumbling action inside the radiator, produced by turbulence and flow separation. Schmued later wrote about the fix for this problem:[21]

> The British Air Ministry was extremely helpful. Among others, they sent Dr. B.S. Shenstone (who arrived February 25, 1941), to assist us in some of the airflow problems into the radiator. The radiator, as we had it, consisted primarily of a fairing, which started at the bottom of the fuselage and enclosed the radiator. Dr. Shenstone advised us to provide an upper lip on the radiator housing, which was about 1-1/2 inches below the fuselage contour. By doing this, we got a much better pressure distribution in the air scoop.

The Engine

The forward part of the streamlined fuselage of the NA-73 was designed around an Allison liquid-cooled engine with a small frontal area, an advantage of such in-line engines compared to air-cooled radial engines with much larger frontal area. The engine, designated the V-1710–39, was rated at 1,120 horsepower at sea level. It had a gear-driven single-stage supercharger that maintained this power level to 11,300 feet, above which the full throttle power dropped off with increasing altitude. The basic Allison V-1710 engine was not new technology; it was first developed by General Motors in 1932, and became a favorite of the US Army Air Corp throughout the later 1930s, finding application on the Curtiss P-40 and the Lockheed P-38. When Schmued designed the prototype NA-73, this engine was a natural choice, being the best available American in-line engine.

The first P-51s off the production line quite naturally went to England, the British having paid for the design, development, and initial production of the airplane. The initial contract called for 400 airplanes. In service with the RAF, these Allison-powered Mustangs performed well at low altitudes but at the 15,000 to 20,000-foot altitudes at which much of the combat with the Luftwaffe took place, their performance was lacking. As a result, the RAF soon relegated the early P-51s mainly to ground attack roles.

[21] Ibid, p. 61.

FIGURE 6.8. North American P-51B Mustang.
(National Air and Space Museum, NASM-7A35577.)

R. J. Mitchell's Spitfire, on the other hand, had been designed around the Rolls-Royce Merlin engine, which was responsible in part for the success of the Spitfire. Because of this, Rolls-Royce in Derby took it upon themselves to modify one of the early NA-73s with a supercharged Merlin engine. In Schmued's words:[22]

> The airplane performed marvelously. As a matter of fact, the British invited Americans to witness the trials of this airplane. The U.S. Air Force was convinced that the airplane would make an excellent air superiority fighter if it had the Merlin engine. So it was agreed between the British Purchasing Commission and the United States Air Force to order the P-51B model, which was a redesign of the airplane.
>
> (Note: Schmued's words were written in the 1970s, hence his use of the current "US Air Force" rather than the World War II Army Air Force.)

On November 30, test pilot Bob Chilton flew North American's first Merlin-powered Mustang, the P-51B. The flight lasted only forty-five minutes because of engine overheating problems. The problem was solved, and flight testing resumed later in December. The performance of the P-51B was markedly better than the earlier Allison-powered aircraft, with an increase in service ceiling of 10,000 feet and an increase in maximum speed of 50 mph at 30,000 feet. Anticipating this result as early as August 1940, arrangements were made for the production of the Merlin in the United States by the Packard Motor Car Company. Production was initiated in September 1940. By the end of the war, twenty-three different models of the Merlin would be produced by Packard. Starting with the P-51B (Figure 6.8), all subsequent Mustangs were powered by the Packard-built Merlins.

[22] Ibid, p. 106–107.

In conceptual airplane design, the choice of an engine is usually made as part of the overall iteration process between weight and sizing, and using the performance analysis to answer the question: Does the design meet the specifications? In the case of the Mustang, a major specification was speed – to design the fastest possible fighter. For Edgar Schmued, this defined the choice of the engine – the most powerful engine available in the United States that would fit the size and streamlined shape of the airplane. These were the considerations that initially lead to the choice of the Allison in-line V-1710 12-cylinder engine that powered the early NA-73s supplied to the RAF. At the time, the Merlin was not even a choice, being foreign made, and with Rolls-Royce straining to provide the number required just by the RAF. It took the special demonstration by Rolls-Royce of the tremendous increase in performance of the Mustang when powered by the Merlin to almost immediately result in the US production of the Merlin engine by Packard. Schmued, as the Mustang designer, did not initially play a role in this development but he certainly embraced the Merlin in all subsequent Mustang versions, starting with the P51B. There is no doubt that the adoption of the Merlin was an example of the inclusion of the best available technology in the Mustang design, just as it was in R. J. Mitchell's design of the Spitfire.

Comment on Conceptual Design and New Technology

The use of the NACA laminar flow airfoil and the Merlin engine in the P-51 are examples of new technology being used *after* the conceptual design process is finished. This in no way compromises the conceptual design methodology established by Barnwell; his methodology was used by Schmued to craft the overall shape and size of the P-51. In earlier airplanes, such as the Douglas DC-3, new technology was incorporated directly into the conceptual design because that new technology existed at the time the design process was initiated, and Art Raymond and his staff were well aware of its existence. This was not the case with the design of the P-51. When Schmued carried out his conceptual design, the Meredith radiator and its use on the Spitfire were not fully appreciated, and the NACA laminar airfoil experiments were classified with very restricted knowledge of their existence. Only afterward were Schmued and his design team aware of this new technology. But that did not stop them from aggressively incorporating it when it did become known to them. In the next chapter we will see even more graphic examples of new technology being incorporated in an airplane design after the conceptual design process is completed. But still, even in the last half of the twentieth century, the conceptual design methodology followed that originally set forth by Barnwell in 1916.

Controversy over Aerodynamics

The laminar flow wing and the Meredith radiator were two high technology features used on the P-51. How much did each one contribute to the superior high-speed performance of the Mustang? Controversy over this question surfaced long after the fact, in the 1990s, between two very different contributors to the P-51 design and development – Ed Horkey, the North American aerodynamicist who was working in the trenches daily with the P-51, and John Leland Atwood, at the time chief engineer and vice president of North American, who along with President Dutch Kindelberger oversaw the successful purchase of the first 400 Mustangs by the British, and later shepherded the acceptance and massive production orders for the P-51 by the US Army. Both men had an engineering education. Atwood graduated from the University of Texas with a bachelor's degree in civil engineering in 1928; Horkey received a bachelor's degree in mechanical and aeronautical engineering from the California Institute of Technology in 1938. By 1940, when the P-51 was designed, Atwood was 36 years old, and had been the chief designer for the famous twin-engined B-25 Mitchell bomber, among other earlier North American airplanes. Horkey was 24 years old when he worked on the P-51 aerodynamics but although still young, he had gained experience working at the Guggenheim Aeronautical Wind Tunnel while at Caltech. Horkey was a strong proponent of adopting a laminar flow airfoil for the Mustang; indeed, as noted earlier, he was the one who took a wind tunnel model of the laminar flow wing to the University of Washington for the definitive tests that resulted in the final wing design of the Mustang.

Well into retirement, Atwood published a paper in 1996 that stated categorically that the Meredith radiator design was responsible for the Mustang's superior performance.[23] With North American's adaptation of the Meredith concept, he noted that the cooling drag was estimated at only 3 percent of the total drag, requiring only 40 horsepower for cooling. In contrast, at the same flight velocity the 400 mph Spitfire with its Meredith radiator required 200 horsepower for cooling. The difference,

[23] Atwood, J. Leland, "Mustang Margin: A Reappraisal," *Air Power History*, Fall 1996, pp. 46–49.

wrote Atwood, was in North American's effective duct system for the radiator which captured the energy recovery that was available in the Meredith concept.

In the same year, also well into retirement, Ed Horkey published his view that "the low drag (laminar flow) airfoil first used on the P-51 was a large contributor to the excellent performance and is still finding good applications today."[24] Horkey then takes a swipe at some of his previous management: "Some authors today, including some management at NAA who were in no way aerodynamicists, have claimed that manufacturing tolerances, slipstream effects, etc., prevented significant laminar flow and low drag." He references some wind-tunnel and flight tests carried out by the NACA to counter these critics. In regard to the radiator design, he makes no value judgment as to the degree of its contribution to the P-51 performance. He simply states that "the cooling radiator and duct were derived by the aerodynamics group in close cooperation with Ed Schmued and his designers." Horkey praised his aerodynamic group, saying that they "were well-educated in aero/thermodynamics." Taking a different approach to the analysis of the Meredith radiator, using calculus instead of algebra, Horkey simply states that "the mathematical work of a British professor (Meredith) was never used by us at any time." As to the relatively far aft location of the radiator, Horkey states that "locating the radiator aft was an overall fuselage/aerodynamic choice, heavily influenced by the simple fact that the radiator size, computed by us and power plant people in preliminary design, was too big to fit forward under the engine." Unfortunately, Horkey was not able to make further contributions to this controversy; he died on July 28, 1996 as a result of serious injuries in a car accident. He was 80 years old.

Between the two, Lee Atwood had the last word in this controversy. In a short two-page editorial in *Flight Journal* in June 1999, he addressed the NACA results for the laminar flow airfoil:[25]

A 1939 report by Eastman Jacobs and others at Langley contained the results of the tests of some small laminar-flow airfoils. The drag on these small models was quite low, and there was some hope that laminar flow could be achieved much father back on an airfoil than had been predicted by previous investigators. The publishers of the report, however, warned that they had not been able to obtain

[24] Horkey, Edward, "The P-51: The Real Story," *Journal of the American Aviation Historical Society*, Vol. 41, No.3, Fall, 1996, pp. 178–189.

[25] Atwood, J. Leland, "An Engineer's Perspective on the Mustang," *Flight Journal*, June 1999, pp. 113–114.

laminar flow on wings of anywhere near the size of those required for actual aircraft and that their tests were to be taken only as the results from laminar-flow models of not more than six inches in width.

Atwood, in a critical tone, then takes a swipe at Horkey, writing:

> In spite of this warning, however, both Ed Horkey (leading aerodynamicist at North American) and Bell Aircraft's chief engineer, Robert Woods, decided to try laminar-flow profiles on the P-51 and the Bell P-63 respectively. These airfoils were incorporated on the Mustang and the Bell airplane with the hope that laminar flow could be extended well back on their wings. Extensive efforts were made to polish and protect the P-63 wing's leading edge profile, but the results were equivocal. Those who advocated the laminar flow wing felt that the Mustang's outstanding performance resulted from laminar flow over most of the wing.

Then Atwood categorically states:

> With respect to the Mustang, many tests – including some in recent years – have shown that extensive laminar flow was *not* developed on the Mustang wing and that the drag of the wing was probably no less than that of conventional wings of the same thickness and taper ratio. On the other hand, the Mustang's cooling drag was much lower.

Atwood then reiterates that the Mustang's exceptional performance was due to the Meredith radiator, and not the laminar flow wing.

About the time this editorial went to press, Lee Atwood died on March 5, 1999, at the age of 94.

We know today that laminar flow usually does *not* prevail over the NACA laminar-flow airfoils used on full-scale aircraft fabricated by conventional manufacturing methods and operating under ordinary field conditions. This is not due, however, to the large size of real airplane wings. Lednicer's[26] modern computational fluid dynamic (CFD) calculations of the flow over a complete full-size P-51 show a large region of laminar flow over the wing. The mathematical boundary condition at the wing surface for these CFD calculations is, however, an infinitely smooth surface. The reality is that the manufactured wings come out of the factory with wavy, rough surfaces, and that nicks, bug spots, etc. encountered at the surface during real operating conditions create even more surface roughness. A turbulent boundary layer is almost guaranteed on a rough surface. It is likely that the P-51 wings in service did not generate large regions of laminar flow. Some hard evidence for this situation was

[26] Lednicer, David, "World War II Fighter Aerodynamics," *Sport Aviation*, January 1999, pp. 85–91.

provided by secret experiments carried out by the NACA during World War II on a Mustang in flight. Using a standard velocity rake mounted on the wing, the measurements indicated that boundary layer transition from laminar to turbulent flow occurred about 15 percent of the distance downstream of the leading edge; i.e., 85 percent of the wing experienced turbulent flow. Atwood's arguments were most likely correct. The NACA laminar flow airfoils, however, yielded a beneficial result totally unexpected by their designers. These airfoils were designed to have the point of minimum pressure on the surface occurring much further downstream of the leading edge compared to conventional airfoils. This gives a much larger region on the wing over which the pressure is decreasing – a much larger region of favorable pressure gradient. This is desirable to try to maintain laminar flow. Having the point of minimum pressure much further downstream of the leading edge also gives the airfoil a higher critical Mach number, allowing the airplane to fly faster before encountering the adverse compressibility effects associated with speeds near the speed of sound. The NACA laminar flow airfoils, therefore, found extensive use on high-speed jet aircraft after World War II. Indeed, there are many high-speed airplanes flying today that still use an NACA laminar flow airfoil section in order to obtain a higher critical Mach number. This is a good example of serendipity in aeronautical design. If anything, it was this airfoil with the higher critical Mach number and not laminar flow that enabled the remarkable high-speed performance of the P-51.

EDGAR SCHMUED – LATER YEARS

At the end of World War II, the production line for the P-51 came to an end. North American, along with the rest of the airplane industry in the United States, found itself in a rapid downsizing and belt-tightening mode. It was also the beginning of the jet age, and North American joined this age with a blockbuster jet fighter design – the F-86 Sabre. This airplane started life on November 22, 1944, as company design study RD-1265. The study was of a high performance jet with a straight wing, initially earmarked for the US Navy. It became the Navy's FJ-1 Fury. On May 18, 1945, the Air Force joined in with an authorization for three prototypes designated the XP-86. However, wind-tunnel tests of the straight wing XP-86 indicated an aerodynamic drag high enough that the airplane could not reach the 600 mph speed dictated by the Air Force. Fortunately, at this point Larry P. Greene, the project aerodynamicist, became aware of German research on swept wings at high subsonic

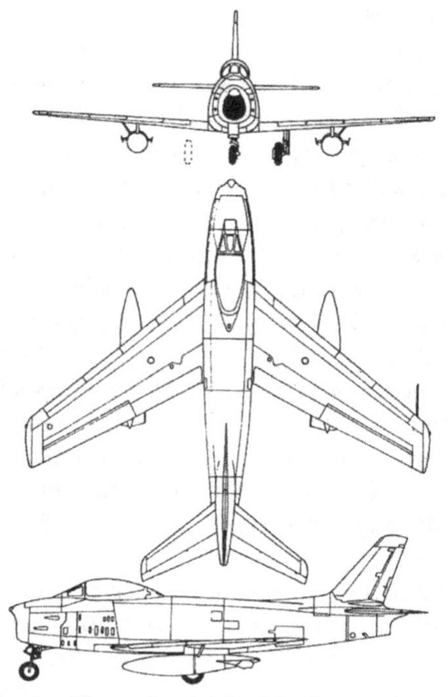

FIGURE 6.9. Three-view of the North American F-86H.

speeds. Along with Ed Horkey, who by now was North American's chief aerodynamicist, Greene decided to use the German swept wing data to redesign the wing of the XP-86, giving the airplane an aesthetically beautiful appearance with a 35 degree swept wing (as best seen in the three-view in Figure 6.9), and most importantly a large increase in maximum speed. On September 15, 1948, one of the first production F-86As set a new official world's speed record of 671 mph. The F-86 Sabre became the Air Force's premier jet fighter during the 1950s, and played a pivotal role during the Korean War, clearing the skies of North Korean and Chinese aircraft and insuring air superiority for the United Nations forces in that conflict. It is interesting to note that the airfoil shape chosen for the F-86 was not a laminar flow section, but rather a thin standard NACA four-digit symmetric airfoil, tapering from an NACA 0009.5 at the wing root to an NACA 0008.5 at the tip. This was an interesting juxtaposition of the ultra-new swept wing technology with a tried and proven airfoil from the middle 1930s.

North American quickly followed with the design of a new jet fighter, the F-100 Super Sabre, the Air Force's first supersonic fighter. By this time, however, Edgar Schmued had less and less to do with the details of design. He was elevated to assistant chief engineer for design, under Raymond Rice, chief engineer. In this position, all new North American airplane designs passed through the hands of Schmued but the details were handled by chief designer Howard Evans. In the postwar period, the engineering manpower required for the design of a given airplane skyrocketed commensurate with the rapidly advancing technology incorporated with such designs. For example, in 1940 about 6,000 man-hours went into the aerodynamic design of the P-51; for the F-86, the aerodynamicists spent about 73,000 hours. Wind-tunnel tests for the F-86 took twenty times longer than for the P-51. By this time, however, North American had its own wind tunnel and was less reliant on organizations such as the NACA and the Guggenheim wind tunnel at Caltech. Constructed in 1942, the North American tunnel created a speed of 325 mph in a 7.75 by 10-foot rectangular test section.[27] In this environment, new conceptual airplane design became a product of teams of people rather than the genius of a single designer. The dominant role played by Schmued in the design of the P-51 was not repeated to the same extent in North American's new jet fighter. Although Schmued is credited by some to be the father of the F-86, in reality, the aerodynamic design of that airplane was due to the project aerodynamicist, Larry P. Greene. As time progressed, the individual airplane designer inside Schmued was feeling more and more uncomfortable.

On August 1, 1952, Edgar Schmued resigned from North American, after twenty-two years with the company. He was 52 years of age, and received no pension. At the time, he offered no real reason for his leaving. Indeed, he was much respected by virtually all his colleagues and his retirement party is remembered as "the biggest ever held for a company executive."[28]

At the time I resigned from NAA, Mr. Kindelburger's health appeared to be failing, and I faced the possibility of seeing the Atwood-Rice team taking over the management. Their adverse attitude towards me made it mandatory to find employment elsewhere. At the time I had made suggestions to improve the combat capabilities of the F-100. These were turned down and it gave me the opportunity

[27] "North American's New Wind Tunnel," *Aero Digest*, Vol. 41, No.2, August 1942, p. 193.

[28] Wagner, p. 191.

to resign from NAA with a legitimate reason. (The U.S Air Force later demanded the changes I had proposed.)[29]

In passing, we note that the F-100, the last airplane that Schmued touched at North American, was also the last successful fighter produced by the company, although a new design, the YF-107A, exceeded Mach 2, but was not ordered by the Air Force in lieu of the new Republic F-105.

Edgar Schmued was subsequently recruited by Northrop Aircraft in Hawthorne, California, and was hired as vice president in charge of engineering. There he oversaw the design of the Northrop F-5, which was eventually purchased by over thirty foreign air forces, but not by the United States. Instead, the US Air Force bought the spin-off trainer version, the T-38. A total of 1,187 T-38s were produced, and it still remains the only supersonic trainer in the Air Force. In 1957, Schmued retired from full-time management at Northrop, remaining as a part-time technical consultant.

Schmued filled the rest of his days serving on advisory boards, working at home as a self-employed technical consultant, and traveling with his third wife, Christel. On June 1, 1985, he died of heart failure in Oceanside, California. Over 250 people attended his memorial service held in a hangar at the Van Nuys airport. One of the eulogists, Buddy Joffrion, made perhaps the most important observation, namely that "The P-51 was Ed's greatest pride and that the fighter that contributed so much to ending the war in Europe was designed by a man born in Germany, powered by an engine designed in Britain, and built on American soil!"[30]

THE THEME CONTINUED

The case history of Edgar Schmued reinforces the theme evolving in this book, namely that the exponential improvement in airplane design in the twentieth century is not due to an exponential improvement in the intellectual methodology of conceptual airplane design, but rather to the willingness of airplane designers to incorporate advanced but proven new technology in their designs, either during or after the conceptual design process. Edgar Schmued is in the ranks of the grand designers because he conceived the P-51, which in turn set new standards for fighter performance during World War II. He went out on a limb by using a newly researched NACA laminar flow airfoil on the P-51 wing, and he recognized the value of the concept of

[29] Letter from Edgar Schmued to M. Stanton, May 8, 1984. [30] Wagner, pp. 224–225.

FIGURE 6.10. Edgar Schmued exiting a P-51.
(National Air and Space Museum, SI-83-6832.)

the Meredith radiator enough to immediately incorporate it in the design. On the other hand, outside of the wind-tunnel tests carried out at Caltech and then the University of Washington used to choose the laminar flow wing over the more conventional wing, virtually no other wind-tunnel testing was used during the design of the P-51. In this, Schmued was of the same mind as R. J. Mitchell, who used no wind-tunnel tests for his design of the Spitfire. During the 1930s, wind tunnels were rapidly improving, and data from them was rapidly becoming more reliable. Nevertheless, airplane designers are as a whole conservative, and as late as the 1930s they were reluctant to base their designs on data obtained in wind tunnels, although as we have seen, the grand designers were eager to embrace new but proven technology; they just did not want to hang their hat on wind-tunnel data. The sands were rapidly shifting however. As discussed earlier, Ed Warner in his 1936 book on airplane design, pointed out the growing reliability of wind-tunnel data and urged airplane designers to make good use of this tool. World War II helped to bring airplane designers closer to the wind tunnel. In fact, after the end of World War II, airplane designers, still conservative, were making reliable wind-tunnel data an essential part of their repertoire. The days when the likes of Mitchell and Schmued could rely primarily on their exceptional "sixth sense" as to what the lines of a successful airplane should be were rapidly disappearing.

7

Defying the Limits

The Jet Airplane, the Second Design Revolution, and Kelly Johnson

It was and is important for an engineer to keep up with advancing technology.

In design, you are forced to develop unusual solutions to unusual problems.

Kelly Johnson, *Kelly: More than My Share of it All*

The scene: the northern edge of the 44 square mile Rogers Dry Lake at Muroc in the Mojave Desert in Southern California. In the middle of nowhere stand a few isolated structures – a small hangar, water tower, and a military barracks that much later would become the massive and famous US Air Force Flight Test Center at Edwards Air Force Base. *The time*: 9:10 am on January 8, 1944. *The action*: a sleek, beautiful airplane, painted green and grey and polished to a high gloss, stands alone on the hard surface of the dry lake, save for a rather excited small group of Army and Lockheed Aircraft personnel making last minute adjustments to the airplane, and generally milling around with great anticipation. A few others are standing on top of a small hill a few hundred yards away, waiting for history to be made. Most of them are wearing jackets and coats to protect against the early desert cold. In the cockpit is Lockheed test pilot Milo Burcham, who has just started the engine. But unlike all conventional aircraft at that time, there is no thundering, vibrating roar from a piston engine, and strikingly, there is no rotating propeller – just a wailing scream from rotating machinery inside the fuselage. Burcham releases the brakes, and slowly the airplane accelerates across the lake bed. At a speed of 110 mph it leaves the ground and smoothly begins to climb. Five minutes later, however, Burcham is back on the ground, the landing gear having

FIGURE 7.1. (a) The Lockheed XP-80 "Lula Bell."
(National Air and Space Museum, SI-75-4844.)
(b) The XP-80 after restoration by the National Air and Space Museum.
(National Air and Space Museum, SI-81-3815.)

failed to retract. It was a minor problem, quickly fixed. By 10:00 am the airplane is back in the air. The Lockheed engineers have affectionately named the airplane *Lula-Belle*, and quickly *Lula-Belle* and Burcham are out of sight. But not for long; suddenly *Lula-Belle* reappears in front of the group, streaking by at low level and at full power with a speed of 475 mph. Burcham pulls up in a steep climb, making a series of rolls in both directions. The airplane flies beautifully – and this on its first flight.[1,2]

Thus, America's first practical jet fighter, the Lockheed XP-80, came on the scene. (Figure 7.1.) Its designer, Kelly Johnson, was there to

[1] The early history of the jet revolution is nicely told by James O. Young in *Lighting the Flame: The Turbojet Revolution Comes to America*, Air Force Flight Test Center History Office, Edwards AFB, California, 2002.

[2] Wooldrige, E. T., Jr., *The P-80 Shooting Star: The Evolution of a Jet Fighter*, National Air and Space Museum, Smithsonian Press, Washington, DC, 1979.

(b)

FIGURE 7.1. *(cont.)*

oversee the first flight. (Johnson was present at the first flights of all his many airplanes throughout his lifetime.) Johnson can be seen in the photograph in Figure 7.2; he is the man in the overcoat shaking the hand of Milo Burcham after the completion of the second flight of the Lula-Belle.

A month later the XP-80 became the first American aircraft to exceed 500 mph in level flight (502 mph at 20,480 feet). A long series of P-80 "Shooting Stars" (later to be redesignated the F-80) followed, the first jet-propelled airplane to serve in quantity in the US Air Force. Although the P-80 never saw combat at the end of World War II, the F-80 was America's first jet fighter to go into action at the beginning of the Korean War. An F-80 shot down a North Korean MIG-15 on November 7, 1950, in that war's first jet combat. Over 1,700 F-80s were produced, the last delivery being June 1950.

When he designed this aesthetically beautiful airplane (Figure 7.3), Kelly Johnson was at the young age of 34 and was already Lockheed's

FIGURE 7.2. Kelly Johnson shaking the hand of test pilot Milo Burcham after the successful second flight of the XP-80, January 8, 1944.
(National Air and Space Museum, NASM-7A48650.)

chief research engineer. He had joined the company in 1933 after receiving a master's degree in aeronautical engineering from the University of Michigan. He immediately participated in the design of the Lockheed Electra series of twin-engines, twin-tail commercial airplanes in the 1930s, and Lockheed's first all-metal aircraft. (Amelia Earhart was flying an Electra when she was lost over the Pacific on July 2, 1937.) He went on to design the Lockheed P-38 Lightning, a twin-boom fighter that became famous during World War II. Just prior to the beginning of the war, he designed the four-engine triple-tail Constellation commercial transport that found great success with the airlines and the public immediately after the war. Johnson's entry into the jet age, the XP-80, was designed and built in just 143 days, a remarkable accomplishment by this young aeronautical engineer. Later, he went on to design such pioneering aircraft as the F-104, the first airplane capable of sustained flight at Mach 2, the U-2 high altitude reconnaissance vehicle, and the stunning SR-71 Blackbird, a unique supersonic Mach 3 (plus) airplane with performance

FIGURE 7.3. The Lockheed P-80 Shooting Star.
(National Air and Space Museum, USAF-31314AC.)

that has yet to be exceeded by any other manned airplane to the present time. Kelly Johnson is a "grand designer" in every sense of the word; some might argue the *grandest* of the grand designers on the strength of the sheer number of his different innovative airplane designs that went on to great success and that paced the discipline of airplane design for years to come. This chapter tells his story. This chapter also serves as the final case history in this book because it gives a modern reinforcement of the overall theme of the book, namely that the exponential advancement in airplane design during the twentieth century was not due to any corresponding advancement in the intellectual methodology of conceptual airplane design, but rather is due to the willingness of talented airplane designers such as Kelly Johnson to incorporate modern but somewhat proven technology in their designs.

When the *Lula-Belle* streaked by the group of Lockheed and Air Force engineers that morning of January 8, 1944, it represented the Second Design Revolution for both Lockheed and Kelly Johnson – the Jet

FIGURE 7.4. The Heinkel He-178, 1939.
(National Air and Space Museum, SI-84-10658.)

revolution. (Unlike the first design revolution that lasted throughout the 1930s, we are still witnessing aspects of the second design revolution today, although it has been morphing from a mantra of flying higher and faster to one of jet airplanes flying more efficiently and in a more environmentally friendly way. The quest for speed and altitude is still with us, however, with new concepts for hypersonic airplanes powered by supersonic combustion ramjet engines – Scramjets; many people, including myself, feel that hypersonic flight is the wave of the future.) The jet revolution, in reality, began in the 1930s, well before Lockheed and Johnson got on board with it, and its roots were clearly in Europe. We start with this story.

THE SECOND DESIGN REVOLUTION – THE JET AGE

On August 27, 1939, a small airplane took off from the Heinkel company airfield in Germany. It had no visible means of propulsion – no external propeller to generate thrust. (Figure 7.4.) Rather it was powered by a nominal 1,000 pound thrust, a gaseous hydrogen-fueled jet engine conceived and designed by Hans von Ohain. With test pilot Erich Warsitz at the controls, the He-178 flew for the first time. Von Ohain's engine on this flight produced 838 pounds of thrust, and pushed the He-178 to a

FIGURE 7.5. The Gloster E28/39, 1941.
(National Air and Space Museum, SI-80-3824.)

maximum speed of 360 mph, in spite of the fact that Worsitz was unable to retract the landing gear, which had locked in the down position. In the words of von Ohain:[3]

> This was the first flight of a turbojet aircraft in the world. It demonstrated not only the feasibility of jet propulsion, but also several characteristics that had been doubted by many opponents of turbojet propulsion: (1) The flying engine had a very favorable ratio of net power output to engine weight – about 2 to 3 times better than the best propeller/piston engines of equal thrust power. (2) The combustion chambers could be made small enough to fit in the engine envelope and could have a wide operational range from start to high altitude and from low to high speed flight.

With this flight of the He-178, the era of the jet-propelled airplane began. The He-178 was purely experimental, but it jump-started the second design revolution – the jet revolution. (Just three days later, Hitler's troops marched into Poland, starting the Second World War.)

More than a year and a half later, on May 15, 1941, another small airplane with no obvious means of thrust took off for the first time from Cranwell in England. The Gloster E28/39 (Figure 7.5) was powered by a turbojet engine designed by Frank Whittle, a young RAF officer. The engine produced 860 pounds of thrust, propelling the E28/39 to a

[3] von Ohain, Hans, in the foreword to *Elements of Propulsion*, 2nd edition, by Jack D. Mattingly, American Institute of Aeronautics and Astronautics, Reston, VA, 2006, p. XXXV.

maximum speed of 338 mph. It was the first flight of a jet-propelled airplane for the Allies in World War II.

At the time neither von Ohain nor Whittle knew of each other's work. Whittle had filed a patent in England in 1930 with a drawing showing his seminal idea for an aircraft jet engine incorporating a two-stage axial compressor followed by a single-stage centrifugal compressor, with the compressed air burned in a ring of combustors, after which it expanded through a turbine, driving the compressors, and then exiting through a ring of nozzles. The compressor and the turbine were connected by a single shaft. Five years later in Germany, von Ohain worked out a patent for a more complete axial flow engine with the air passing axially through a compressor, burner, turbine, and a central nozzle. In the subsequent patent search, von Ohain and his patent attorney could not find any similar patent. It was not until 1937, when the German Patent Office held Whittle's patent against some of von Ohain's claims that von Ohain found out about Whittle's early work. The issue of patents aside, during the war both Germany and England vigorously pursued the development of the jet engine under great secrecy. For decades afterwards, debate raged as to who gets credit for the invention of the jet engine. Both von Ohain and Whittle subsequently moved to the United States, von Ohain to Wright Field in Dayton, Ohio, and Whittle to the Naval Academy in Annapolis, Maryland. The debate was finally put to rest when both Whittle and von Ohain agreed among themselves to share equally in the credit for the invention of the jet engine.

The turbojet engine created a second design revolution, and allowed aeronautical engineers to design new airplanes that approached, and then eventually exceeded, the speed of sound. The speed limitations of propeller-driven aircraft were well understood by the mid-1930s. For example, T. P. Wright, vice president for engineering at the Curtiss Airplane Company, stepped forward at the height of the first design revolution and delivered a bold paper on December 29, 1936, to the Aeronautic Meeting of the American Association for the Advancement of Science at Atlantic City on the subject of "Speed – and Airplane Possibilities." In his paper, Wright presented a graph showing speed versus chronological year for airplanes, reproduced here as Figure 7.6. The solid curves are reality for the period before 1936. The dashed curves are Wright's guess for the ensuing ten years and all of the dashed curves bend and approach limits that are at 500 mph or slower. Of course, all the aircraft considered were propeller-driven. Wright considered these speed

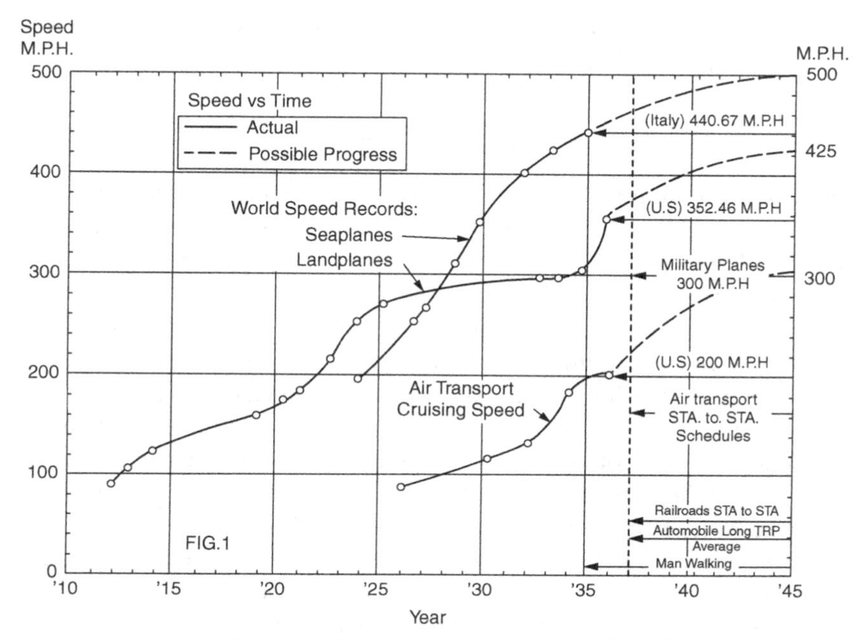

FIGURE 7.6. Speed of aircraft versus years, plotted by T. P. Wright in 1936.

plateaus to represent a "probable ultimate limit." To the present, no propeller-driven airplane has ever broken the speed of sound.

Shock waves are the culprit. The airspeed over the tip of the propeller, due to the combination of its rotational speed with the forward speed of the airplane, can sometimes exceed the speed of sound, resulting in shock waves at the tip and a consequent dramatic loss of propeller efficiency. This problem was completely eliminated by the dispensing of propellers altogether with the advent of the turbojet engine.

Shockwaves, however, became a serious problem for the whole aircraft when the powerful turbojet engines pushed airplanes to speeds around and above Mach one. The high air pressure behind shock waves, impressed on the forward-facing parts of the airplane, greatly increased the pressure drag of the airplane, particularly on the nose, the leading edges of the wings and tails, and on any other protruding parts. This increase in pressure drag is called *wave drag*. Aerodynamic design during the first design revolution was dominated by streamlining to reduce pressure drag due to flow separation; aerodynamic design during the second design revolution was, and still is, dominated by efforts to reduce wave drag. For flight at any speed, a way to reduce wave drag is to reduce the strength of the shock waves, and this is accomplished by innovative tailoring of the airplane shape.

THE SWEPT WING

The first such innovation was the use of a swept wing. The concept was first introduced by the German engineer Adolf Busemann in 1935 during the fifth Volta Conference held in Rome. The topic chosen for the conference was "High Velocities in Aviation." Participation was by invitation only, but it included some of the most famous aerodynamicists from around the world. Busemann conceived the swept wing for use on supersonic airplanes, but it soon became obvious to the Germans that a swept wing would allow a high-speed subsonic airplane to fly close to the speed of sound before encountering the large increase in drag associated with the transonic flight regime. (Swept wings allowed higher drag-divergence Mach numbers.) The German Luftwaffe classified the swept wing concept in 1936, one year after the conference. In contrast, the aerodynamicists attending the conference from other countries, including Britain and the United States, paid little attention to Busemann's idea. Germany set into motion a research program on swept wings that produced a mass of technical data through the end of World War II, data that was uncovered much to the surprise and concern of the Allied technical teams that swooped into the German research laboratories in early 1945. Independently, and without knowledge of Busemann's work, R. T. Jones, the leading NACA aerodynamicist at that time, showed theoretically the advantage of swept wings for high-speed flight. The combination of Jones' theory and the discovery of the German data transformed the airplane design community in the United States. For example, the North American XP-86, originally designed to be a straight-wing jet fighter, was quickly changed to a swept-wing configuration shown in Figure 6.9. Similarly, the Boeing B-47, originally designed to be a straight-wing jet bomber, was quickly changed to a swept-wing configuration. Figure 7.7 illustrates the design evolution of the B-47, showing the adoption of a swept wing in September 1945. From that time on, swept wings have become a familiar design feature of the second design revolution.

THE OTHER ALTERNATIVE

For supersonic airplanes, swept wings became one of two alternative design features to reduce supersonic wave drag. The other was to employ

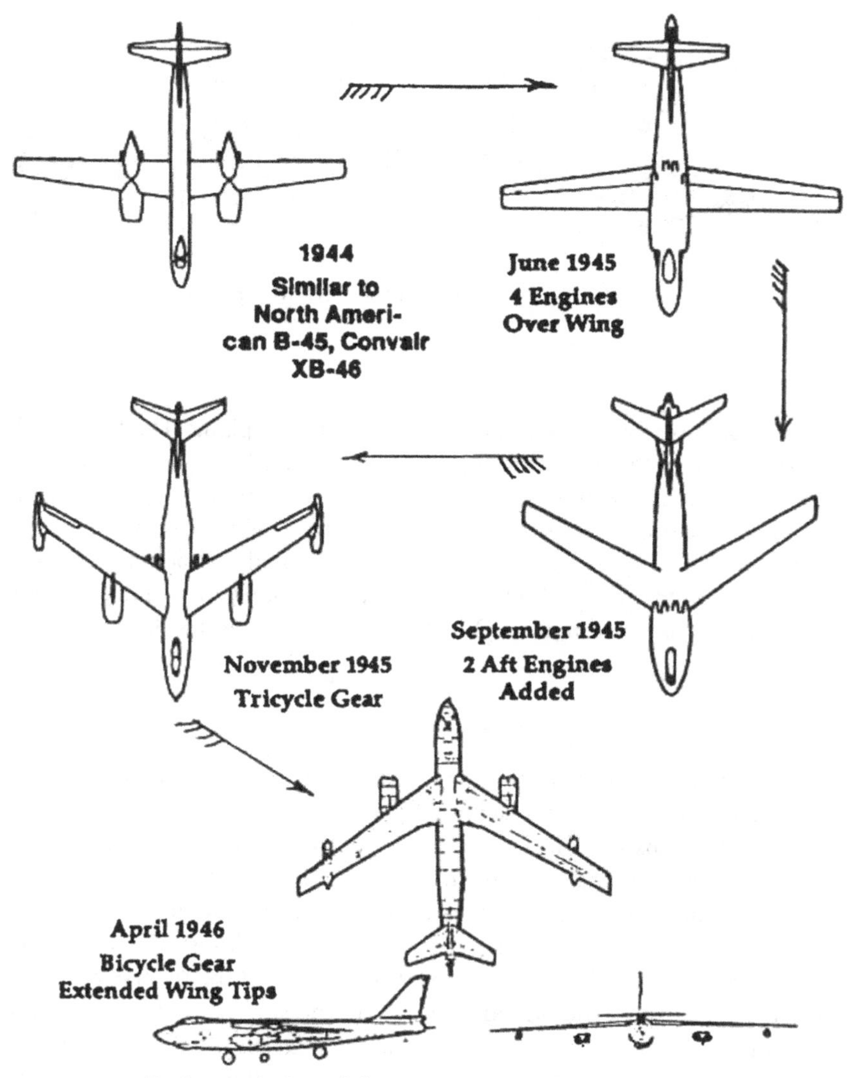

FIGURE 7.7. Design evolution of the B-47. (W. H. Cook, *The Road to the 707*, TYC Publishers, Bellevue, Washington, 1991.)

a small aspect ratio straight wing (short and stubby) with a very thin airfoil. A short, stubby wing has a short leading edge that provides a minimum of frontal area for the high pressure behind the leading edge shock wave to press on. The very thin airfoil with a sharp leading edge weakens the leading edge shock wave. Combined, these two design

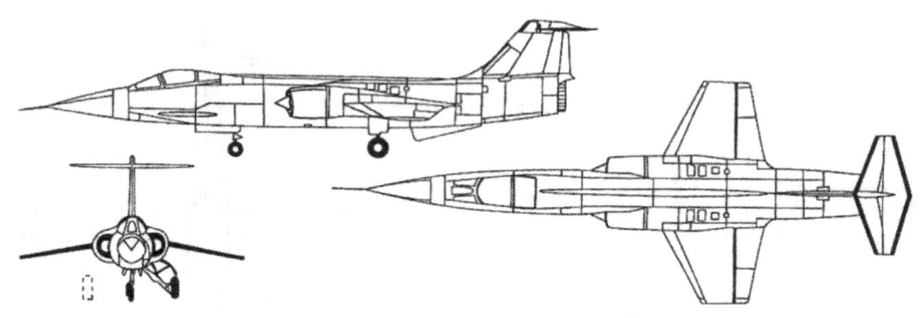

FIGURE 7.8. Three-view of the Lockheed F-104.

features can greatly reduce the wave drag for the airplane. The Lockheed F-104, shown in three-view in Figure 7.8, is an excellent example of good supersonic design, with its sharp-nosed and slender fuselage, a wing with a very low aspect ratio of 2.45, and an airfoil with only 3.4 percent thickness, all designed to reduce the strength of the shock waves and hence reduce the wave drag. The F-104 was the first airplane to be designed for sustained flight at Mach 2. Its designer was Kelly Johnson.

THE AREA RULE

For an airplane to fly at supersonic speeds, it must first accelerate through the transonic flight regime centered around Mach one. As the flight velocity of a subsonic airplane approaches the speed of sound, pockets of locally supersonic flow occur in expansion regions over the wing and around the fuselage. These pockets of locally supersonic flow are usually terminated by a nearly normal shock wave. The almost discontinuous jump in pressure across the shock creates a rather severe adverse pressure gradient that causes the boundary layer on the surface to separate from the surfaces in the vicinity of the shock impingement, hence causing an attendant increase in the pressure drag due to flow separation. This combination of the existence of the shock wave itself (with the attendant reduction of the useful energy of the flow across the shock – in thermodynamic terms, an increase in the entropy of the flow), and the flow separation caused by the severe adverse pressure gradient at the point of shock impingement on the surface, creates a huge increase in drag called *transonic drag divergence* that must be overcome for the airplane to penetrate into the supersonic region. This large increase in drag near

Mach one led to the myth of a "sound barrier" that appeared in the popular literature at the end of the 1930s.

In 1951, Convair ran smack into this problem. The company had designed one of the "century series" fighters intended to fly at supersonic speeds. Designated the YF-102 (Figure 7.9a), the aircraft had a delta-wing configuration and was powered by the Pratt and Whitney J-57 turbojet, the most powerful engine in the United States at that time. On October 24, 1953, flight tests of the YF-102 began at Muroc Air Force Base, and a production line was being set up at Convair's San Diego, California, plant. However, as the flight tests progressed, it became painfully clear that the YF-102 could not fly faster than sound; the transonic drag rise was simply too great for even the powerful J-57 engine to overcome.[4]

Independently, and at almost the same time, Richard Whitcomb, an aerodynamicist at the NACA Langley Memorial Aeronautical Laboratory, had intuitively reasoned that the peak of the transonic drag rise for an airplane could be reduced by designing the configuration with a smooth variation of cross section area from nose to tail. After all, artillery shells routinely passed through the transonic flight regime, leaving the gun muzzle at supersonic speeds, and artillery shells had smooth variations of cross section area along their axes. For an airplane, however, to ensure a smooth variation of cross-sectional area at the location where extra area is contributed by the wings and tail, the fuselage cross-section area had to be necked down, giving the airplane a wasp-like, or "coke bottle" shape. Whitcomb ran a series of definitive tests in the Langley 8-Foot High-Speed Tunnel, which had been rebuilt with a slotted-wall test section to achieve stable transonic flow, proving the validity of his intuition. Thus was born the *transonic area rule*, one of the most important aerodynamic breakthroughs of the second design revolution. After consultation with the NACA aerodynamicists and inspection of the area-rule findings obtained at Langley, the Convair engineers modified the YF-102 to become the YF-102A, with an area-ruled fuselage (Figure 7.9b). On December 20, 1954, the prototype YF-104A left the ground at Lindbergh Field, San Diego, and exceeded the speed of sound while still climbing. The use of the area rule had increased the top speed of the airplane by 25 percent. The production line rolled, and 870 F-102As were built for the US Air Force. The area rule made its debut in dramatic style.

[4] Anderson, John D., Jr., *The Airplane: A History of Its Technology*, American Institute of Aeronautics and Astronautics, Reston, VA, 2002, p. 319.

(a)

(b)

FIGURE 7.9. (a) Convair YF-102 (top) and (b) Convair F-102A (bottom). Illustration of the area rule.
(National Air and Space Museum, SI-89-11936.)

In addition to the advent of the swept wing and the area rule, another aerodynamic breakthrough was the supercritical airfoil, also the brainchild of Richard Whitcomb. For conventional airfoils, once the critical Mach number (the freestream Mach number at which sonic flow is first encountered at some point on the airfoil) is exceeded, the drag-divergence Mach number occurs just slightly above this. Whitcomb found that with proper shaping of the airfoil, essentially flattening the top surface, the drag-divergence Mach number was increased well beyond the critical Mach number. Such airfoils flew in a type of "grace period" regime where a much larger increase in free stream Mach number above the critical value was tolerated by the flow before drag-divergence occurred – hence the term "supercritical" airfoils. (Technical comment: the relatively flat upper surface of the supercritical airfoil tends to reduce the lift of the airfoil, but this is compensated for by reflexing the tailing edge of the airfoil to increase the lift.)

ADVANCES IN PROPULSION

The invention of the jet engine in the 1930s, with its subsequent use for the first jet-propelled airplanes in the 1940s, was a revolution on a par with the invention of the first practical airplane by the Wright brothers. Both Frank Whittle's and Hans von Ohain's early designs, however, were centrifugal flow jet engines – engines where the airflow coming into the engine is funneled into the center of large compressor blades, which then sling and accelerate the air into a diffuser. In the diffuser, the outward flowing air is then turned and slowed, achieving the desired pressure increase across the compressor before entering the combustor. Some of the early production jet engines, such as the Allison J33 (a direct descendant of the Whittle design) that powered the Lockheed P-80 jet fighter, were centrifugal-flow jet engines. These engines, however, presented a large frontal area, making them less desirable for installation in the sleek, streamlined airframes demanded by high-speed airplanes. As the jet revolution progressed, centrifugal-flow jet engines receded to the background, replaced by the more slender axial-flow engines. In these engines, the airflow is compressed by passing through an alternating series of rotating blades (the rotor) and stationary blades (the stator), where the overall flow path is essentially along the axis of the engine – hence the label "axial flow."[5]

[5] Ibid, pp. 334–335.

A raw comparison between a propeller and a turbojet engine is that a propeller produces less thrust but with higher efficiency, and a turbojet produces more thrust with less efficiency. The advantages of both the propeller and the turbojet were merged in the late 1950s with the advent of the turbofan. Here, a turbojet engine serves as a core that produces power to drive a large fan made up of a large number of blades acting in the same manner as a propeller; turbofan engines produce high thrust with higher efficiency than a turbojet. Today, the three major aircraft engine companies – Pratt and Whitney, GE, and Rolls-Royce – manufacture an extensive line of turbofan engines. Some of these are large engines with thrust levels over 100,000 pounds. The bypass ratio, defined as the amount of air passing through the fan divided by the amount of air passing through the core, has increased from about 1.5 for the early turbofans to over 9 for modern engines. Although turbojet engines are still manufactured, mainly for application to supersonic airplanes, the term "jet engine" today conjures up a turbofan. Since the Wright brothers' first successful flight in 1903, the holy grail of aeronautical engineers has been the design of airplanes to fly faster, higher, and further. In this chapter we will see a graphic example, the SR-71 "Blackbird" designed by Kelly Johnson and his Skunk Works team. Flying faster than Mach 3 at altitudes above 80,000 feet, the SR-71 was powered by two Pratt and Whitney J-58 bleed bypass turbojet engines, especially designed for use on this airplane. For flight at Mach 3 and higher, the engine must be highly integrated with the airframe. The J-58 engine core itself produced only about 17 percent of the total thrust at Mach 3. The rest of the thrust was produced by the aerodynamics of the nacelle, which changed in flight to make it work more like a ramjet as speed increased. Such engine/airframe integration is a hallmark of the second design revolution.

Today, the second design revolution is characterized by an explosion of new ideas, concepts and airplane designs. For example, aeronautical engineers are questing after dramatic 60 to 70 percent improvements in the efficiency of commercial airplanes, looking ahead thirty to forty years in the future. (As mentioned earlier, this is an example of the jet revolution partially morphing into a quest for improved efficiency, which might be labeled a third design revolution, the "efficiency revolution." But since it still involves jet propulsion. this author prefers to consider it part of the extended context of the second design revolution.).. Another example is the goal of sustained hypersonic flight in the atmosphere for both military and commercial purposes. This quest is driving new research on supersonic combustion ramjet engines – Scramjets – which appear to be the only viable propulsion system for such applications.

All this is just a taste of the extended second design revolution. It is much more complex and extensive than the first design revolution discussed in Chapter 4, and it is still ongoing.[6] Straddling both the first and second design revolution is Kelly Johnson, arguably the grandest of the "grand designers." This chapter tells his story and examines his philosophy and technique of airplane design. In his lifetime, Johnson designed more iconic and different airplanes incorporating a wider spectrum of the existing modern technology than any other airplane designer. Moreover, these airplanes were immensely successful; many of them set the gold standard for their class. Johnson not only did not hesitate to hang new technology on his innovative designs, but when such technology did not exist, he developed it. Kelly Johnson's design philosophy is the capstone for the theme presented in this book. Johnson did not revolutionize the intellectual methodology used for conceptual airplane design; rather, he applied new technology in ways different from anybody else to produce revolutionary airplanes. His story begins in Northern Michigan at the beginning of the twentieth century.

KELLY JOHNSON – THE FORMATIVE YEARS (1910–1940)

Clarence Leonard Johnson was born on February 27, 1910 in Ishpeming, Michigan, a small mining town near Marquette. He was the seventh of nine children born to Peter and Christine Johnson, Swedish immigrants. Peter came to the United States in 1882 from the small city of Malmo, Sweden, seeking a better life. He had saved $600 to buy a farm in Nebraska, and was making his way across the country when in Chicago he was bilked out of his money by a group of swindlers. Thinking they put him on a train to go to his new farm in Nebraska, Peter instead found himself in Northern Michigan. He got off at Marquette and found work laying railroad ties. A mason in Sweden, he was later hired as a bricklayer by a local construction company. Two years later he had earned enough money to send for his intended bride, Christine Anderson, in Sweden. After marrying, they settled in a rented house in Ishpeming.

The Johnson family was, as Kelly put it in his autobiography, "poor but not in spirit."[7] His father would work whenever the harsh Northern

[6] Ibid. See Chapter 7, pp. 283–359. This chapter gives a concise overview of the second design revolution and its impact on airplane design.

[7] Johnson, C. L., with Maggie Smith, *Kelly: More Than My Share of it All*, Smithsonian Institution Press, Washington, DC, 1985, p. 1.

Michigan weather would allow; he could lay over 2,000 bricks a day. With the help of the older daughters, Christine took in washing for additional family income. Even Kelly became part of the business, picking up and delivering laundry several times a week.

The name "Kelly" came when Johnson was in second grade. Cecil, the class bully, would mockingly call Johnson "Clara" for Clarence. Cecil was a foot taller than Kelly, so retribution called for some planning. One recess in the schoolyard, while exchanging words, the kids pushed the two together. Johnson kicked Cecil behind the knee, tripping him, and then jumped on him as he fell. With a loud pop, Cecil's leg was broken. Johnson admitted to his teachers that he had done it, and moreover tripped Cecil on purpose. Kelly stayed home from school the next day but when he returned, the kids welcomed him back with a popular song of the day, "Kelly from the Emerald Isle." The kids named him Kelly, and it remained for a lifetime.

Kelly Johnson credits his parents for his love of learning. Johnson was encouraged to study hard in school and to also read on his own. He *wanted* to go to school, many days arriving early to be first in line. He went almost daily to the public library. He read over and over the Tom Swift series for young men, including *Tom Swift and His Airplane, Tom Swift and His Electric Automobile, Tom Swift and His Submarine* – indeed, *all* the Tom Swift series. (I know firsthand how impressive this series of books can be on the mind of an 8-year-old boy.) This love of learning stayed with Kelly Johnson for the rest of his life and is an underlying reason for his phenomenal success as an airplane designer, as we will see later in this chapter.

Kelly Johnson knew from the age of 12 that he wanted to be an airplane designer. Much later he wrote that "my whole life from that time was aimed at preparing for that goal."[8] Focused and studious, he earned good marks in school. He built hundreds of model airplanes, read extensively on aeronautics, and compiled a book of clippings on aircraft. His motivation came not from any experience in flying in airplanes at the time, but rather from the love of what they were (flying machines). He was drawn by an almost aesthetic attraction to these machines, with an intellectual desire to understand how and why they flew. The year 1922 was just four years after the end of World War I. Most airplanes in use were surplus holdovers from the war, and the market for new

[8] Ibid, p. 7.

airplane designs was sparse. Nevertheless, 12 year old Kelly Johnson had a vision of what he was going to do for the rest of his life.

The building industry in Michigan was improving and in 1923, Peter Johnson moved his family to Flint where Peter was able to do well in the construction business. The Flint public school system was excellent; flourishing in classes there, Kelly was an A student, taking every opportunity to write and talk about aviation. Having saved $350 from helping his father in construction, he was tempted to take time off after his high school graduation and work his way around the world on a boat. He was dissuaded by one of his high school teachers, however, who explained to Kelly the value of continuing directly to college. So he entered Flint Junior College instead, where he studied physics, general mathematics, and calculus, and earned extra money as a tutor in calculus.

THE FORGING OF AN AERONAUTICAL ENGINEER: THE UNIVERSITY OF MICHIGAN

In the southeast corner of Michigan, near Detroit, one of the most prestigious public universities in the United States was flourishing. The University of Michigan was founded in 1817 as a territorially chartered college in Detroit, a frontier town at the time. In 1837, the university moved to its present location at Ann Arbor. In 1913, a series of lectures on aeronautics was given at the university by Felix Pawlowski and Lucien Marchis. Pawlowski was a Polish engineer who in 1913 had worked with the Russian aviation pioneer and airplane designer Igor Sikorsky on the four-engine *Russkiy Vityaz* (Russian knight), later named the Grand. The Grand was the world's first four-engine airplane, a gargantuan machine that established the reputation of Sikorsky as an important airplane designer before World War I. Prior to that, Pawlowski had taken the first course in aeronautical engineering ever given, by Leucian Marchis in 1910 at the University of Paris, and had worked with Gustav Eiffel on Eiffel's wind tunnel in the fashionable Paris suburb of Auteuil.[9] Thus, when Pawlowski and Marchis were invited to give lectures at the University of Michigan in 1913, it was quite a feather in the university's hat to have two such distinguished European aeronautical engineers on campus. Marchis returned to the University of Paris, but Pawlowski was offered a teaching assistantship in mechanical engineering in 1913 and spent the

[9] Anderson, John D. Jr., *A History of Aerodynamics*, Cambridge University Press, Cambridge, England, 1998, p. 269.

rest of his professional life at Ann Arbor. In 1914, promoted to instructor, Pawlowski introduced Michigan's first course in aeronautical engineering, Theory of Aviation. By 1915, this had developed into the first four-year curriculum in aeronautical engineering at any American university. At the time, this four-year bachelor's degree program was part of the Department of Naval Architecture, Marine Engineering, and Aeronautics.

Pawlowski carried the European tradition of research to Michigan. In the 1920s, when the embers of aeronautical research were barely glowing in the United States, fanned slightly by the research program of the National Advisory Committee for Aeronautics (NACA) at the Langley Memorial Aeronautical Laboratory in Hampton, Virginia, Pawlowski pushed to build an aeronautical research program at Michigan. He saw to it that a major wind tunnel was included in plans to build the East Engineering Building on campus. Started in 1924, the wind tunnel was built into the base of the building. At the time, it was the largest wind tunnel at any American university with a test section measuring 8 feet from side to side. The maximum airspeed was 80 mph, sufficient for carrying out meaningful tests on various model shapes, including models of whole airplanes.

In 1926, the head of one of the wealthiest families in America, Daniel Guggenheim, established The Daniel Guggenheim Fund for the Promotion of Aeronautics, with a grand sum of $2,500,000.[10] One of the four major undertakings of the fund, as Daniel Guggenheim wrote to Commerce Secretary Herbert Hoover at the time, was "to promote aeronautical education in both institutions of learning and among the general public."[11] One of the seven universities ultimately receiving Guggenheim sponsorship was the University of Michigan. Awarded $78,000 by the Fund, the University used $28,000 for the completion of its wind tunnel and new equipment, and the remainder to establish for ten years a Daniel Guggenheim Chair of Aeronautics. The first occupant of the Chair was Lawrence Kerber, a Michigan graduate who was in charge of the aerodynamic section of the US Army Corps research center at the McCook Field in Dayton. However, two years later Kerber resigned the Chair to take a position with the Commerce Department, and in 1929 Felix Pawlowski became the new Guggenheim Professor at Michigan.

[10] Hallion, Richard P., *Legacy of Flight: The Guggenheim Contribution to American Aviation*, University of Washington Press, Seattle, Washington, 1977.

[11] Ibid, p. 45.

With Pawlowski on the faculty was Herbert Sadler and like Pawlowski, he was European. After teaching marine engineering at the University of Glasgow, Sadler joined the University of Michigan at the turn of the century as a professor of naval architecture and marine engineering. Sadler had an inherent interest in aviation; his great-granduncle James Sadler was England's first balloonist, making an ascent on May 5, 1785.[12] When Pawlowski showed up at Michigan, Sadler joined forces with him to help establish the aeronautical engineering program in 1915. Also in the early 1920s, Edward Stalker joined the faculty as an assistant professor. Stalker was an early graduate of Michigan's aeronautical engineering program, receiving his B.S.Ae.E in 1919 and his master's degree in 1923. And there was Ralph Upson, with a mechanical engineering degree from the Stevens Institute of Technology, who joined the Michigan faculty as a leading expert in lighter-than-air vehicle design. In the words of the distinguished aeronautical historian Richard P. Hallion: "By the mid-1920s, then, the University of Michigan had an advanced department of aeronautics taught by members of unusual ability."[13] It is no surprise that Michigan, along with the likes of MIT and Caltech, was chosen as one of the seven universities to receive grants from the Daniel Guggenheim Fund for the Promotion of Aeronautics.

In 1930, the University of Michigan's aeronautical engineering program became a separate department of aeronautical engineering, with Edward Stalker as chair of the department. One year earlier, in 1929, Kelly Johnson became a student at Michigan. He could not have fallen into a better place. Because he was a Michigan resident with limited financial resources, the campus at Ann Arbor was a logical choice. The mating of one of the best programs in aeronautical engineering with Johnson's steadfast desire to become an airplane designer was synergistic to both parties involved. Kelly was excited. "But the real beauty of it for me when I enrolled in 1929 was the distinguished faculty," he later wrote. "Many had impressive national, even international, reputations in their fields. I thought I never could be so smart as these men. I couldn't wait to begin my classes."[14]

Felix Pawlowski was Johnson's professor for the first course in aerodynamics; he also hired Kelly to work for him in the wind tunnel, which involved the design of a Union Pacific streamlined train, a smoke removal project for the city of Chicago, and a very early proposal for generating

[12] Ibid, p. 54. [13] Ibid, p. 54.
[14] Johnson, Kelly: *More Than My Share of It All*, p. 14.

energy with a wind machine. Johnson was getting some early, valuable wind-tunnel experience.

By 1930, Kelly's excellent academic progress was noted by Ed Stalker, who was Johnson's advisor for his course of study. Stalker hired Kelly as a student assistant, and immediately put him to work in the wind tunnel with some degree of autonomy. Realizing that there were a lot of vacant hours in the wind-tunnel schedule, and being an enterprising soul, Kelly obtained permission from Stalker to rent out some of the vacant time for thirty-five dollars a day plus power charges. Joining with his best friend in college and fellow aeronautical engineering student, Don Palmer, they both became part-time proprietors of the wind tunnel, basically pocketing the rental fee with Stalker's permission. They aggressively approached potential customers. One of these was the Studebaker Motor Company who hired them to test the Pierce Silver Arrow. Johnson and Palmer understood the nature of pressure drag due to flow separation; this was a major source of drag on the large ugly headlights on the Pierce Silver Arrow, eating up 16 percent of the power at 65 mph. Convinced, the Studebaker designers shaped the headlights into the fenders. The Pierce Silver Arrow ultimately became one of the early totally streamlined cars. With this and many customer jobs in the wind tunnel, both Johnson and Palmer became rather well-to-do based on student standards. They were eventually stopped, however, when the faculty became aware of just how lucrative this extracurricular business became.[15]

Later in life, Kelly Johnson always considered himself to be an aerodynamicist as well as an airplane designer. This had a lot to do with his excellent education in aerodynamics at Michigan. Edward Stalker published *Principles of Flight: A College Test on Aeronautical Engineering*[16] in 1931, right in the middle of Johnson's undergraduate years at Michigan. As I read this book today against the backdrop of modern classes in aerodynamics, it is clear that Stalker's book provided the 1930s era Michigan students with an amazingly sophisticated and for its day, a modern perspective in aerodynamics. Stalker, for example, gives a complete discussion of Prandtl's lifting-line theory for calculating the induced drag and the reduction of lift encountered by a finite wing. Although

[15] Adamson, T. C., Jr., "Aeronautical and Aerospace Engineering at the University of Michigan," *Aerospace Engineering Education During the First Century of Flight*, American Institute of Aeronautics and Astronautics, Reston, Virginia, Chapter 4, p. 49.

[16] Stalker, Edward, *Principles of Flight: A College Text on Aeronautical Engineering*, Ronald Press Co., New York, 1931.

Prandtl had worked through his lifting-line theory by the end of World War I, almost a decade passed before it became common knowledge in the English-speaking countries. Even then most American universities considered such material to be appropriate for graduate level courses only, reflecting a heavy emphasis on *practical* undergraduate engineering education in contrast to the engineering science education prevailing in Europe. At the University of Michigan, however, Pawlowski and Sadler carried with them the European tradition in engineering education. The Aeronautical Engineering Department did not hesitate to offer their undergraduates a science-based education, and Kelly Johnson understood and appreciated such an emphasis. Stadler's book was used in the undergraduate program at Michigan. Moreover, Johnson became very familiar with this book, about which he later wrote: "When Professor Stadler published a new text, I had the temerity to write out all the answers to all the problems, and proposed to publish them." Kelly Johnson loved mathematics, and solving all the problems in Stalker's book was natural for him. Johnson went on: "I was persuaded not to do this since it would undermine the book and considerably diminish sales."[17]

In his autobiography,[18] the most noted aerodynamicist of the twentieth century, Theodore von Karman, makes note of the European influence on aeronautical engineering education at Michigan. Professor von Karman, then at Aachen University in Germany, was making an extended visit, his first, to the United States in 1926 at the invitation of the renowned physicist, Dr. Robert Millikan, president of the California Institute of Technology in Pasadena, California. Millikan's ultimate goal was to recruit von Karman to join the faculty at Caltech. It took several additional visits and three more years before von Karman accepted a full-time professorship at Caltech. In 1926, however, as part of his overall orientation to engineering education in the United States on his first trip, von Karman visited Felix Pawlowski in Ann Arbor in October. He also visited New York University and MIT. "At the first two universities," von Karman noted, "I was struck with the dominance of Europeans. At Michigan the professor of aeronautics was a man of Polish origin named Pavlowski, [sic] who was trained in Paris. The head of N.Y.U.'s aeronautics department was Alexander Klemin, an immigrant from London's East End, trained in England. Both these men

[17] Johnson, Kelly: *More Than My Share of It All*, p. 17.

[18] von Karman, Theodore, with Lee Edson, *The Wind and Beyond: Theodore von Karman*, Little, Brown and Co., Boston, 1967, p. 127.

made very important contributions to education of young American aeronautical engineers." One of these young American aeronautical engineers was Kelly Johnson.

The year 1932 was not a great year to graduate from college and look for a job. The country was in the throes of the Great Depression. It took Kelly only three years to finish his bachelor's degree, and he and his classmate Don Palmer, driving a car borrowed from one of their professors, set out for California. Nobody in the aircraft industry had an engineering job to offer either of them. The only light on the horizon for Johnson and Palmer, however, was provided by Richard von Hake, soon to be production manager of Lockheed. In spite of producing successful aircraft made of plywood, such as the Lockheed Vega, the company had gone into receivership. Just a month before Johnson and Palmer appeared on the scene at Burbank, the company with assets of $129,961 and in a totally rundown condition, was snapped up by the stockbroker and banker Robert Ellsworth Gross for only $40,000. Gross was not familiar with the aviation industry but he had sound business sense. He was to save the company. Looking to a better future for Lockheed, von Hake suggested to the two young Michigan graduates: "Look, something is going to come of this. Why don't you go back to school and come out again next year. I think we'll have something for you."[19] The two returned to Michigan to work on master's degrees.

The extra year at Michigan was important to Kelly Johnson's future career for two reasons:

(1) The obvious benefit of a graduate education, which challenged his intellect and expanded his knowledge of aeronautical engineering. He was awarded the Sheehan Fellowship which paid his expenses for the year. At the graduate level, it is necessary to pick a field of specialization; Kelly chose both the supercharging of engines to obtain high power at high altitude (engines were a strong interest of his advisor, Ed Stalker), and boundary layer control, an important aspect of aerodynamics for the reduction of aerodynamic drag. About this choice, Johnson later wrote: "I always loved engines and aerodynamics. It was a natural choice, as well."[20]

(2) Both Johnson and Palmer were again assigned to work in the University of Michigan wind tunnel. As a side project, they designed and tested automobile shapes for racing in the famed

[19] Johnson, Kelly: *More Than My Share of It All*, p. 19. [20] Ibid, p. 20.

> Indianapolis Memorial Day event, receiving notice in the local newspaper:[21] "Five of the qualifying cars which will race at Indianapolis Memorial Day have bodies designed by two University graduate students, C. L. Johnson and E. D. Palmer. All of the cars are semi-stock Studebakers and all qualified for the race at speeds ranging between 110 and 116 miles per hour." Johnson and Palmer were once again up to their old undergraduate ploy, using the wind tunnel to do work generated by themselves. Of much more importance, however, was a contract between Michigan and the Lockheed Aircraft Corporation to test in the wind tunnel a model of the new twin-engined transport designed by Lockheed. Designated the "Electra," this airplane was the outcome of Lockheed's new reorganization; the company was depending on this new airplane to pull it out of financial ruin. In fact, the Electra is what prompted von Hake's optimism in encouraging both Johnson and Palmer to come back after a year in graduate school. "I think we'll have something for you." So here was Kelly Johnson, a University of Michigan graduate student, carrying out wind-tunnel tests under the supervision of the well-known and respected Professor Ed Stalker, on a model of an airplane on which Lockheed was depending for its very survival.

During the course of these tests, Johnson noticed that the airplane had stability problems; both the longitudinal and directional stability was "very bad" in his judgment. The opinion of a graduate student, however smart, did not carry the same authority as that of Ed Stalker, who in consultation with Lloyd Stearman, the first president of the reconstituted Lockheed Aircraft Corporation, decided the wind-tunnel data was acceptable. The wind-tunnel report was finalized and submitted to Lockheed.

Kelly Johnson received his master's degree from Michigan in 1933. In the 1930s, having a college degree in aeronautical engineering was rare, and having a graduate degree rarer still. This was especially the case for designers of airplanes, who tended to be conservative practitioners of a trade rather than academically inclined thinkers who were as much interested in the "why" as in the "how" of things. The grand designers considered in this book were a spotty lot in regard to higher education. The Wright brothers never attended college. Frank Barnwell earned his bachelor's degree at the University of Glasgow by going to night school. Art

[21] Ibid, p. 20

Raymond had a master's degree from MIT under the tutelage of Ed Warner. R. J. Mitchell was self-taught, never attending college. Edgar Schmued was also self-taught. Kelly Johnson, however, was of a new generation. He was a resident of the state of Michigan which, to his good fortune, had one of the best degree programs in aeronautical engineering in the country, and because the University of Michigan was a state school, Johnson had the wherewithal to attend. Moreover, he was academically inclined – he *liked* going to school, and he made good grades. Graduating with a master's degree in aeronautical engineering from the University of Michigan under the tutelage of the respected Ed Stalker gave him a special cachet. (Attending college in the 1920s and 30s, especially to study aeronautical engineering, was in all fairness, beyond the reach of many aspiring airplane designers, both in the United States and in Europe. First, there was the cost, beyond the resources of many young people especially in the deep Depression era. And then were was the matter of admission requirements, and having the requisite academic background and family pedigree to be admitted to such schools as MIT in the United States, and Cambridge and Oxford in England. Take R. J. Mitchell for example. With working-class parents and education at the local Hanley High School, R. J. had no image of attending the likes of Oxford or Cambridge; at the age of 16 Mitchell left high school and his father entered him as an apprentice with a locomotive engineering firm. In sharp contrast was the engineering education of Nevil Shute, the noted author. Like Mitchell, Shute was highly motivated by airplanes as a young boy. But unlike Mitchell, Shute's family background – his father was a high placed official in the London Post Office – allowed him to obtain ready admittance to Oxford as an engineering student immediately after World War I. From there, Shute went on to work for de Havilland, then Vickers, and finally to found Airspeed, Ltd. In 1930, that later produced large numbers of successful small twin-engine transports through the end of World War II, some of which Shute had a hand in designing. The world knows Nevile Shute as a famous and successful author of novels, but few people realize that he spent the first two-thirds of his life as an aeronautical engineer and airplane designer – but not in the panoply of the grand designers. Clearly, a college education did not necessarily spawn a grand designer, nor did the lack of a college education preclude the emergence of a grand designer.)

The summer of 1933 found Kelly Johnson and Don Palmer on the road again, driving to California in search of jobs in the aeronautical industry. This time, however, they met with success. Johnson was hired by Lockheed in Burbank, and Palmer by the Vultee Airplane Company in Glendale.

FIGURE 7.10. The Loughead brothers' Model G floatplane, 1912. (National Air and Space Museum, NAM-A-45682.)

THE LOCKHEED AIRCRAFT COMPANY AND KELLY JOHNSON IN THE 1930S

The massive Lockheed-Martin Aerospace Company in 2012 celebrated its 100th anniversary by reaching back to the founding brothers, Allen and Malcolm Loughead (pronounced Lock-Heed), who built their first airplane, the Model G floatplane (Figure 7.10), in a garage at Pacific and Polk Streets in San Francisco. Both brothers had earned respect as automobile mechanics, the source of their income. This was not enough, however, to completely finance the building of the flying machine, and they had to obtain funding from several investors. The Model G made it through its early flight trials successfully, but in the summer of 1913 it was damaged. The company went out of business.

Undaunted, in early 1916, Allan and Malcolm founded a second company, the Loughead Aircraft Manufacturing Company. Setting up shop in rented space in a garage in Santa Barbara, the new company hired a 21-year-old architectural draughtsman named John K. "Jack" Northrop, and plunged into the design of a new flying boat. The single prototype flew successfully in March, 1918, but too late to capitalize on large orders from World War I. The company was liquidated in 1921.

Alan Loughead tried again in 1926. Along with Jack Northrop and a former Air Service flying instructor, W. Kenneth Jay, Loughead formed the new Lockheed Aircraft Company (the first time the new spelling of Lockheed was used) principally to produce a new airplane designed by Northrop called the Vega. Production was set up in a rented building in Hollywood. Jack Northrop had both feet firmly planted in the new emerging design revolution and he designed a beautifully streamlined

FIGURE 7.11. Lockheed Vega.
(National Air and Space Museum, SI-2009-7961.

single-engined, cantilever monoplane. The Vega attracted the attention of George Hearst, Jr., publisher of the *San Francisco Examiner*, who purchased the first production model before its construction was finished. The Vega (Figure 7.11) was successful, and Lockheed received more orders, forcing them to move to a larger building located in Burbank.

In March 1928, Jack Northrop, who held the position of chief engineer, and Ken Jay left the company. Gerald F. "Jerry" Vultee became the new chief engineer. Orders for the hand built Vega increased, with production growing from two in 1927 to forty just in the first six months of 1929.[22] The new Lockheed Aircraft Company was booming. Because of this healthy situation, the principal stock holder in Lockheed, the wealthy brick and tile manufacturer Fred Keeler, saw a business opportunity to make a large profit by selling the company. With Allan Loughead protesting, Keeler sold Lockheed in July 1929 to the Detroit Aircraft Corporation. Three months later, the stock market crashed, sending the United States into the Great Depression. The market for new aircraft essentially evaporated.

In July 1929, the Lockheed Aircraft Company became the Lockheed Aircraft Corporation, a division of the Detroit Aircraft Corporation. Although Lockheed continued to produce Vegas as well as derivative Orions and Altairs, its profits were drained by the parent company to cover massive losses. In October, 1931, the Detroit Aircraft Corporation went into receivership, carrying the Lockheed Aircraft Corporation with it. On June 16, 1932, the doors to the factory were locked.

[22] Francillon, Rene, Jr., *Lockheed Aircraft Since 1913*, Pitnam and Company, Ltd., *London, 1982, p. 11.*

But not for long. Just five days later, a new Lockheed Aircraft Corporation, like the Phoenix, rose from the ashes. Led by the 35-year-old investment broker Robert E. Gross, a group of investors paid $40,000 for Lockheed's assets. With an entirely new team in charge, the doors were again opened. The well-known airplane designer Lloyd C. Stearman was installed as president, Robert Gross as treasurer, Richard von Hake as chief engineer, and Hall L. Hibbard as assistant chief engineer. Twenty-two airplanes remained to be completed. But the new managers knew quite well that the future of the company hinged on the design and development of a new aircraft. A new ten-seat, single-engine, all-metal transport had earlier been designed by Stearman and Hibbard. Gross, on the other hand, argued for a twin-engine airplane because, following the wake of the twin-engine Boeing 247 and Douglas DC-1, he felt that twin-engine airplanes were the future for civil transports. (Gross was prescient; two years later the federal government forbade single-engine transports operating in the United States from carrying passengers at night or over land not suitable for emergency landings.) The Lockheed team agreed with Gross and the twin-engine Electra was born.

It was this environment into which Kelly Johnson and Don Palmer fell when they first visited the Lockheed Aircraft Corporation in the summer of 1932. Things could hardly have been worse for the company, but the design of the new Electra instilled optimism and it was with this optimism that von Hake sent the two graduates back to the University of Michigan, encouraging them to return a year later. That year made all the difference. The company was now counting on the new Electra for its survival and Johnson's serendipitous involvement with the wind-tunnel testing of the Electra at Michigan made him that much more interesting to Lockheed. When the company hired him in the summer of 1933, little did they know how strong a role Kelly Johnson was to play in the continued survival of Lockheed.

Johnson's initial contact at Lockheed in 1932, Richard von Hake, had been made production manager of the newly reconstituted company, and Hall L. Hibbard took his place as chief engineer. Hibbard had a master's degree in aeronautical engineering from MIT and was the principal airplane designer of the new Electra. The decision to hire Johnson in 1933 was made jointly by Hibbard and Cyril Chappellet, the assistant to the president and personnel officer. It was not an easy decision because the engineering department of Lockheed at that time consisted only of five engineers and Hibbard was being very careful about whom he added. The tipping point with Hibbard was Johnson's direct involvement with the first set of wind-tunnel tests on the Electra carried out at Michigan, plus Johnson's excellent academic record at the school. Rather than starting in the engineering department, however, Kelly Johnson began his career with Lockheed assigned to the tooling department, designing tools for the manufacture of the Electra. Over the next three months, Johnson, in his own words, "discovered that there was a lot to learn about tooling."[23]

Johnson's continued concern about stability problems with the original Electra design, problems he had observed during the wind-tunnel tests at Michigan, began to weigh on the mind of Hibbard. Finally he sent Johnson back to the Michigan wind tunnel, this time as a Lockheed employee and with a wind-tunnel model of the Electra in the back of Johnson's car.

In 1934, wind-tunnel testing of a complete airplane configuration in its conceptual design stage was still the exception rather than the rule. The community of airplane designers was just beginning to feel a degree of comfort with the reliability of wind-tunnel data. (Recall that R. J. Mitchell intentionally eschewed wind-tunnel testing for the design of the Spitfire.) Lockheed's Hall Hibbard, in contrast, believed in the viability of wind-tunnel testing. It was part of his education at MIT; by the late 1920s the school had three wind tunnels of considerable size with 4-foot, 5-foot, and 7.5-foot diameter test sections. The 4-foot tunnel was modeled after the National Physical Laboratory wind tunnel in Britain and was built at MIT by Jerome Hunsaker. The other two were added by Ed Warner when he was head of the department. Hall Hibbard learned early on to be comfortable with wind tunnels, and that they could be a serious and viable tool as part of the airplane design process. He was happy to use the University of Michigan wind tunnel during the design of the Electra.

The Electra model in the back of Kelly Johnson's car had a single vertical tail and fillets at the juncture of the wing with the fuselage, products of the early conceptual design of the airplane. These features cannot be found in the three-view of the final Electra design seen in Figure 7.12. The wind-tunnel tests that Johnson was about to carry out at Michigan would be his first substantive input to the design of an airplane.

Figure 7.13 shows Kelly Johnson and the Electra model in the University of Michigan wind tunnel. In setting up and carrying out the tests, Johnson was guided by two of the most viable textbooks available at the

[23] Johnson, Kelly: *More Than My Share of It All*, p. 23.

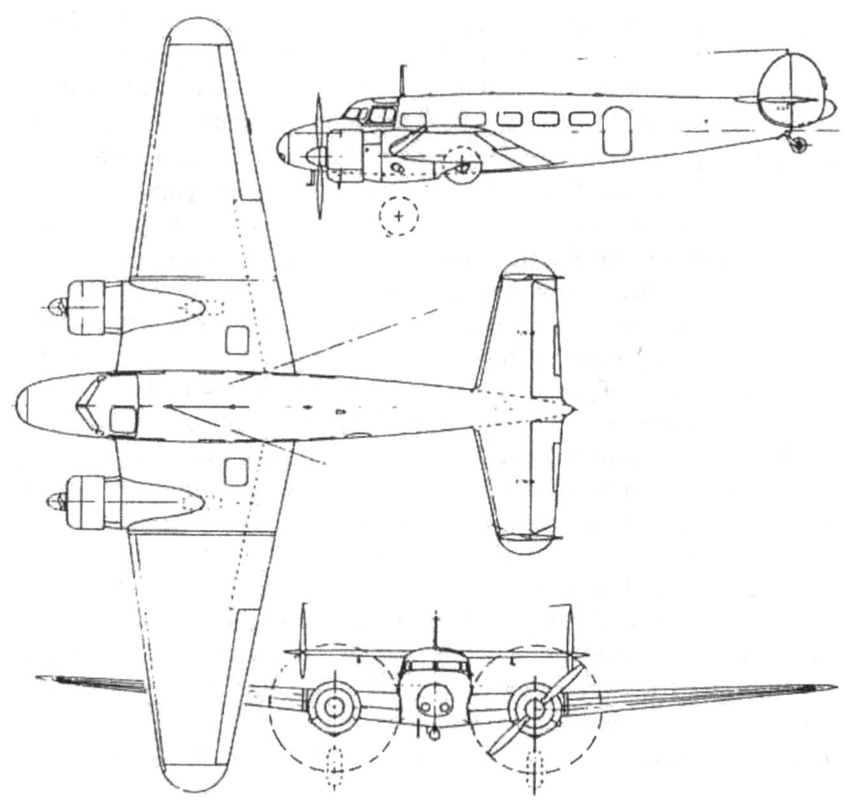

FIGURE 7.12. Three-view of the Lockheed Electra Model 10-A.

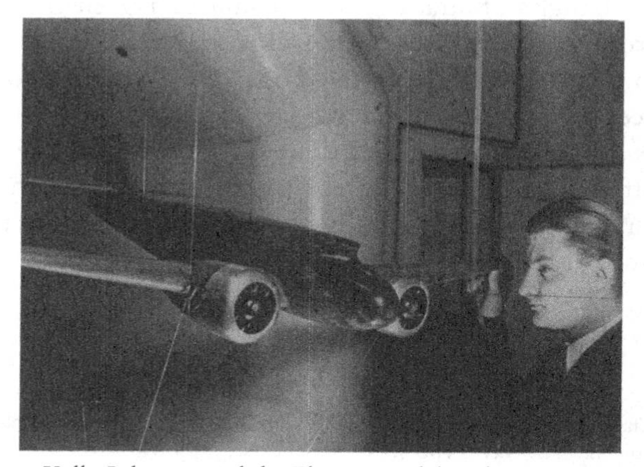

FIGURE 7.13. Kelly Johnson and the Electra model in the University of Michigan wind tunnel, 1934.

time: Edward P. Warner's book *Airplane Design*, 1927 and Ed Stalker's *Principles of Flight*, 1931. As noted earlier, Warner's book was almost a bible going into the design revolution, and Stalker's book was an integral part of Johnson's education at Michigan. Johnson references both books in a paper published one year later by the newly formed Institute of Aeronautical Sciences.[24] This is the first archive journal paper published by Johnson, and it is devoted totally to the seventy-two wind-tunnel runs made by him on the Electra model. Careful study of this paper gives insight into the mature engineering thinking of Johnson, even at this very early stage of his career. Overall, the paper is a model of succinct technical writing. It clearly explains how the wind-tunnel data proved that the initial Electra configuration was longitudinally unstable. Johnson included a graph of the variation of the moment about the center of gravity of the airplane plotted versus the angle of attack. The necessary criteria for longitudinal stability is for the slope of the curve to be negative – a decreasing moment as angle of attack is increased. The data for the original Electra configuration showed a negative slope at low angles of attack, but a sudden reversal to a positive slope above four degrees angle of attack. What this meant to Johnson was the airplane was stable at the low angles of attack for cruising speed, but was unstable at the higher angles of attack for slower speeds, and for takeoff and landing. All the remaining wind-tunnel tests were aimed at solving this problem. The solution speaks volumes about Kelly Johnson's power of engineering acumen.

He found that the engine nacelles were essentially blanketing the flow over the horizontal tail and when the nacelles were removed, "satisfactory stability at all angles of attack"[25] was obtained. The engine nacelles, of course, were a necessary feature of the airplane; one could not just get rid of them. So Johnson then concentrated on the aerodynamic flow over the wing between the nacelle and the fuselage. He found that the turbulent flow in this region was being directed over the tail by the fillets and with the fillets off, Johnson so nicely stated that "the wake missed the tail in the best manner." So the first substantive design change to the Electra was the disappearance of the fillets.

[24] Johnson Clarence, L., "Longitudinal Stability of a Bi-Motor Transport Airplane," *Journal of the Aeronautical Sciences*, Vol. 3, No. 1, Sept. 1935, pp. 1-6.

[25] Ibid, p. 3.

It occurred to the writer that the use of endplates on the horizontal tail would improve the stability by increasing the effective aspect ratio of the surface. As finally developed, these endplates were made to serve as fins and rudders so that the original vertical surface was removed entirely.

Thus, the design of a twin-tail for the Electra was born. Subsequently, twin-tailed airplanes became almost a signature feature of Lockheed designs until the advent of the second design revolution and jet airplanes.

Johnson continually fed the results of his wind tunnel tests to Hibbard in Burbank, who in turn wrote back:[26,27] "You may be sure that there was a big celebration around these parts when we got your wires telling about the new find and how simple the solution really was. It is apparently a rather important discovery and I think it is a fine thing that you should be the one to find out the secret."

The original design of the Electra had been carried out by Hibbard and von Hake in full knowledge of the advancements associated with the design revolution. That is why the Electra was designed as an all-metal airplane with an NACA engine cowl, variable-pitch propellers, and retractable landing gear. Fillets, being a rather new feature introduced during the design revolution and adopted for the new Boeing 247 and Douglas DC-1 transports, were also made part of the Electra design, mainly because it seemed like the right thing to do. However, the main aerodynamic advantage of a fillet is to smooth the flow and inhibit flow separation at the wing-fuselage juncture of a low-wing design with a rounded fuselage, where the wing makes a rather acute angle with the side of the fuselage.[28] The Electra, on the other hand, had a fuselage with an almost rectangular cross section, and the wing, although placed low on the fuselage, made essentially a right angle with the vertical side of the fuselage. Aerodynamically, the Electra did not need fillets. When the fillets were dropped from the Electra design on the basis of Kelly Johnson's wind-tunnel findings and recommendation, there was all gain and no loss.

When Johnson returned to Burbank, he found a redesigned Electra incorporating a twin-tail and having no fillets. He also found that he had been made a full-fledged member of the engineering department, which now numbered six engineers. Hall Hibbard liked Johnson, and his initial favorable intuition about Johnson's engineering ability was now justified.

[26] Ibid, p. 4. [27] Johnson, Kelly: *More Than My Share of It All*, p. 24.
[28] Anderson, John D., Jr., *Introduction to Flight*, 7th edition, McGraw-Hill, New York, 2012, pp. 565–568.

Moreover, the fact that Hibbard felt the wind-tunnel results to be important certainly contributed to Johnson writing and publishing a paper in the *Journal of the Aeronautical Sciences*. The journal editor also appreciated the importance of the paper; it was made the lead article. Although little is made about this paper in later histories of Johnson's life (Kelly Johnson himself does not even mention it in his autobiography), I consider it to be a seminal document in revealing Johnson's mind-set. First of all, he *wanted* to publish. Most airplane designers in the twentieth century did not feel this need or have this desire. Secondly, he was educated at Michigan in an environment where his professors published, thus instilling an appreciation for the value of disseminating information to the technical community. Indeed, such sharing and dissemination was the primary purpose for the creation of the Institute of the Aeronautical Sciences (IAS) in the early 1930s. Johnson simply got on board. He went on to publish five more papers in the *Journal of the Aeronautical Sciences* before World War II.

Although Kelly Johnson did not design the Electra, his name soon became associated with the airplane. He received full credit by the company for the vital design changes that were the product of his wind-tunnel tests. Then he was made the flight test engineer for the initial flight test program for the Electra. The first flight of the Electra took place on February 23, 1934, with only the experienced test pilot Eddie Allen on board. Its success was reported in a local newspaper:[29] "Soaring gracefully into the air on its maiden flight, the sleek all-metal airliner flew easily, making another great stride in commercial speed development of air transportation." For the remainder of the initial flight test program, Kelly Johnson flew as the flight test engineer. Once again Lockheed was placing a great deal of responsibility on this 24-year-old relatively fresh college graduate, a testimonial to the growing recognition of his aeronautical engineering abilities. Of this experience, Johnson later wrote:[30]

> In Eddie Allen, I had an excellent teacher. After the first flight, I flew with him as flight test engineer through the entire initial flight test regime – dive tests, stalls, spins, everything. He taught me what it was all about, what was important, what to record.

Johnson went on to accumulate 2,300 hours as the flight test engineer on a number of Lockheed designs, including nine first flights. From this, he

[29] Johnson, Kelly: *More Than My Share of It All*, p. 26. [30] Ibid, p. 26.

adopted the philosophy "that those who design aircraft also should fly them – to keep a proper perspective."[31]

His experience as a flight test engineer led to Johnson's second archive journal paper, also published in the *Journal of the Aeronautical Sciences*.[32] It dealt with flight testing of the Lockheed Model 12 – a slightly smaller version of the Electra – and involved the determination of the lift and drag characteristics of the airplane measured in flight. The publishing of this paper – Johnson's second archive journal paper in as many years – speaks again to his intellectual nature, to a mind-set different than the conventional airplane designer at the time. Simply stated, Johnson felt compelled to publish.

A second aspect about this paper, its very title, hints at the way Johnson is working at Lockheed, not as an airplane designer (as yet), but rather as a research engineer. The title of the paper is "Flight Test Research on a Small Bi-Motor Airplane," and he is being perceived at Lockheed as their "research" person. Indeed, in 1938 he was made chief research engineer of the Lockheed Aircraft Corporation – still a young man of the age of 28. With this title came the authority to hire a few new engineers, which he did, drawing on students he knew from the University of Michigan. Johnson's research image at the company was not just a result of his master's degree from Michigan. Indeed his boss, Hall Hibbard, also had a master's degree, and from the equally prestigious aeronautical engineering program at MIT. But Hall Hibbard was chief engineer, with all the heavy responsibilities that accompany that title, and he had little opportunity to carry out any work that could be described as research. Kelly Johnson, on the other hand, had the opportunity and the bent to do so. He carried this mind-set with him when he created the Lockheed Skunk Works much later on, and virtually every airplane he designed at the Skunk Works was in many respects a major "research project." Johnson remained Lockheed's chief research engineer until 1952, when he was made chief engineer at Lockheed's Burbank plant, rising to vice president for research and development in 1956. Although Kelly Johnson is known throughout the aerospace profession and by the general public as well, as the man who designed some of the most unique airplanes of the twentieth century, few realize that Johnson considered himself to also be a researcher at heart.

[31] Ibid, p. 27.

[32] Johnson, Clarence L., "Flight Test Research on a Small Bi-Motor Airplane," *Journal of the Aeronautical Sciences*, Vol. 4, No. 11, Sept. 1937, pp. 473–477.

Johnson's 1937 paper also speaks to the matter of wind tunnels. By that time Lockheed, as well as most of the aircraft industry in Southern California, was using the 10-foot tunnel at the California Institute of Technology for model testing. Built at Caltech in the early 1930s, and initially designed by the Caltech professors Clark Millikan and Arthur Klein, with later input from Theodore von Karman, the 10-foot tunnel became a workhorse for the aeronautical industry during the 1930s. (This is the same tunnel used by Douglas for model testing of the DC-1, 2 and 3 series, as described in Chapter 4.) In his 1937 paper, Kelly Johnson refers to extensive wind-tunnel tests in the Caltech tunnel of a one-seventh scale model of the Lockheed Model 12. The Reynolds number in the tests was 1,600,000, about one-tenth of the value for the real airplane in flight at high speed. For these conditions, the drag coefficient measured in the wind tunnel should have been larger than that occurring in free flight. However, in spite of this large difference in Reynolds number (always a problem in wind-tunnel testing of models of real airplanes), Johnson proved that the minimum drag coefficient measured in the wind tunnel and measured in free flight were almost the same, the difference being made up by the added drag of rivets, window recesses, ventilating scoops, and "various other items."[33]

In the mid-1930s, it was not common for airplane companies in the United States to have their own wind tunnels. Such facilities were expensive to build and operate. Besides, it was one of the roles of the NACA to provide state of the art wind tunnels and make them available to industry for testing. Unfortunately for Lockheed and other companies on the West Coast, the NACA wind tunnels at the time were at the only NACA laboratory, namely the Langley Memorial Laboratory, situated on the East Coast at Hampton, Virginia. Clearly, there was a critical need for wind tunnels on the west coast to cater to the growing concentration of the aircraft industry in Southern California. Robert A. Millikan, President of the California Institute of Technology in Pasadena, recognizing the excellent opportunity to fold the interests of the rapidly developing aircraft industry with education and research at Caltech, initiated the construction of the large 10-foot tunnel as part of the newly formed Guggenheim Aeronautical Laboratory at California Institute of Technology (GALCIT). Indeed one of the many incentives for the hiring of von Karman at Caltech was the desire to tap his experience with large wind

[33] Ibid, p. 473.

tunnels at the University of Aachen helping Millikan's son Clark, and Arthur Klein to design and build the tunnel. It became operational in 1928 and by the late 1930s was overwhelmed with demands for testing. Six other aircraft companies in addition to Lockheed were renting time in the tunnel.

So it came about that Kelly Johnson's first complete design project with Lockheed was not an airplane but rather a wind tunnel. Seeing the bottleneck beginning to form with the renting of time in the Caltech wind tunnel, and recognizing that Lockheed could no longer afford to wait in line, in 1939 Johnson went to the Lockheed leadership to propose construction of their own wind tunnel. Not since Frank Barnwell convinced the Bristol management to build a wind tunnel in 1919 (see Chapter 3) had a "grand" airplane designer been instrumental in getting his company to build a wind tunnel. Lockheed management understood the need and gave Johnson $360,000 (over four million in today's dollars) for a wind tunnel. Johnson carried out the aerodynamic design himself and one of his engineers, E. O. Richter, designed the structure. Contracting out the construction, the bare tunnel cost Lockheed $186,000 to build and left Kelly Johnson with the remainder for the latest state of the art instrumentation, and for the model shop. The tunnel test section was an 8 by 12 foot rectangle, large enough for testing sizeable models; along with a maximum test-section airstream velocity of 314 mph, the tunnel provided a higher Reynolds number simulation than the Caltech tunnel. This tunnel proved invaluable to Lockheed for fine-tuning future designs during World War II, particularly the P-38 Lightning.

When Kelly Johnson was hired by Lockheed in 1933, it was not as an airplane designer straight away. He was just one of six engineers who worked on various smaller tasks – on "anything and everything they threw at me" as Johnson later wrote.[34] Indeed, his duties as the flight test engineer for the Electra became his main focus in those early years with Lockheed. However, Hall Hibbard had great expectations for Johnson, and he began to phase Kelly slowly into the airplane design process.

An improved version of the Electra, the Model 14 Super Electra (Figure 7.14), was designed by Hibbard with the help of Johnson in 1935. The Model 14 had two new innovations due to Johnson: the Fowler flap for a high lift device, and slots in the wings for directing airflow over the wings. A Fowler flap sits inside the wing during

[34] Johnson, Kelly: *More Than My Share of It All*, p. 25.

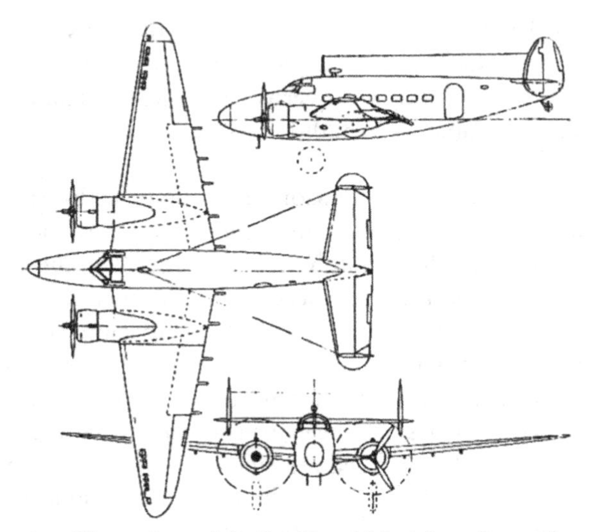

FIGURE 7.14. Three-view of the Lockheed Model 14-Super Electra, 1935.

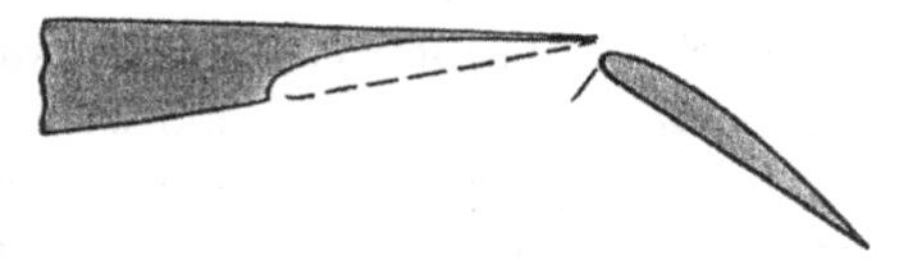

FIGURE 7.15. Fowler flap.

high-speed flight but is deployed out the trailing edge and deflected downward for increased lift at low speeds, as shown in Figure 7.15. In this way both the wing area and camber are increased, resulting in a considerable increase in the maximum lift coefficient of the wing. The Fowler flap was invented by Harlan D. Fowler in 1924. Fowler was an engineer for the US Army at McCook Field in Dayton, Ohio, but his invention was a private venture using his own money.[35] Unable to find financial support for almost a decade, Fowler had some limited success after the NACA ran tests in 1932 that suggested his flap had some value. Glenn Martin in Baltimore hired Fowler to design flaps for several new Martin airplanes, including the Martin 146 bomber in 1935, but these airplanes never went

[35] Anderson, John D., Jr., *A History of Aerodynamics and Its Impact on Flying Machines*, Cambridge University Press, New York, 1998, p. 367.

FIGURE 7.16. Lockheed Hudson, 1939.

into production. It fell to the Lockheed Model 14 to be the first production line airplane to use Fowler flaps. For his work with the design of the Fowler flaps on the Model 14 – he called them Lockheed-Fowler flaps – Johnson received his first major award as an engineer, the Lawrence Sperry Award in 1937 for "important improvements of aeronautical design of high speed commercial aircraft." The slots (Johnson called them "letter box" slots because they resembled the long, thin slots in letter boxes) were placed near the leading edge, and went all the way through the wing from the bottom to the top surfaces. These slots, located in the wing-tip region near the leading edge, can be seen in Figure 7.14 and in the photograph in Figure 7.16. With the wing at moderate to high angle of attack, the higher pressure air on the bottom surface blew through the slots and provided a high energy stream of air that blew over the top surface, energizing the boundary layer over the top surface, thus keeping the flow attached to the wing surface at higher angles of attack and hence delaying the wing stall. This was an effective stall preventive device, and it served as the subject of Johnson's third archive journal paper in the *Journal of the Aeronautical Sciences,* entitled "The Effect of Fixed Wing Slots on the Stall Characteristics of a Modern Bi-Motor Airplane."[36] The paper is focused on extensive flight testing of the Model 14, concentrating on the stall behavior. Once again, as in Johnson's earlier papers, the writing is clear, concise, and informative. Moreover, Johnson includes detailed aerodynamic data measured on the Model 14 of a nature that today would most likely be considered proprietary and not allowed by the

[36] Johnson, Clarence L. and Thoren, Rudolph L., "The Effect of Fixed Slots on the Stall Characteristics of a Modern Bi-Motor Airplane," *Journal of the Aeronautical Sciences*, Vol. 6, No. 11, Sept. 1939, pp. 437–445.

company to be published. The Model 14 was the latest and greatest of Lockheed's airplanes at the time. Its first flight was on July 24, 1937 and Johnson's paper was published two years later with the detailed aerodynamic data on the aerodynamics of its wing and its stall performance. Johnson gives graphic details of the fight test program, the instrumentation (some unique), and the results obtained. He included photographs of tufts on the wing surface that showed the flow over the wing for both the unstalled and stalled conditions.

In the conclusion section of his paper, Johnson reached out to wider implications of his results:[37]

> The foregoing investigation has brought out many factors which will be of great influence on the design of subsequent airplanes. The methods used and the equipment developed have proved very satisfactory in all respects for measuring stall phenomena as well as general stability and control factors. The use of fixed wing slots was thoroughly justified for the case investigated, although it is believed that in the design of a new airplane, certain elements can be embodied which will give equally good stall characteristics at less cost in drag; or the wing slots themselves can be improved in design and arranged to be closed during normal flight so that even an airplane with good stalling characteristics could be improved by the use of the fixed slot unit.

This paper says a lot about Kelly Johnson's professional growth as an aeronautical engineer with Lockheed in the late 1930s. He is showing insight and technical maturity that is propelling him upward in the esteem of Lockheed's technical staff and management. They give him his first opportunity to work on a significant airplane design as a colleague with Hall Hibbard, the experienced chief designer. The Model 14 Super Electra is a new, considerably improved airplane over the Model 10 Electra. Compare the three-view of the Model 14 in Figure 7.14 with that of the Model 10 in Figure 7.12. We see a more streamlined airplane with more highly tapered wings. The airfoil section used on the wing of the Model 10 is the old, tried and proven Clark Y, the 18 percent thick Clark-Y-18 at the root and the 9 percent thick Clark Y-9 at the tip, whereas that used on the Model 14 is a new high performance NACA airfoil, the 18 percent thick NACA 23018 at the root, tapering to the 6 percent thick NACA 23006 at the tip. The NACA five-digit 230-series airfoils were developed and tested by the NACA in 1935, and they produced more lift than the previous NACA four-digit airfoils developed in the early 1930s. (Indeed,

[37] Ibid, p. 445.

some NACA five-digit airfoils are still used today.) These airfoils were considered the best, most modern shapes in the late 1930s and we see the Hibbard/Johnson team readily adopting such a new airfoil for the Model 14.

We also see the Model 14 as a product of a budding, new "grand designer;" a person willing to adopt new but somewhat proven technology in his design. The NACA five-digit airfoil was new but proven technology; so was the Fowler flap. The wing slots were absolutely new, developed by Kelly Johnson and proven by the flight tests described in his 1939 paper.

The Super Electra went on to become Lockheed's most important airplane in the late 1930s. It was designed to compete with the likes of the Douglas DC-3 and Boeing 247. Its top speed was 257 mph and it cruised at 237 mph, 30 mph faster than any other commercial transport at that time. This was mainly due to the high wing loading of 31.8 pounds per square foot for the Model 14 compared to that of 24.3 pounds per square foot for the DC-3. In turn, the Super Electra's high wing loading demanded a highly effective high-lift device in order to have a safe landing speed, and this was the main driver for the use of the Fowler flap. Ultimately, 112 Model 14s were produced by Lockheed, and an additional 119 in Japan by Tachikawa and Kawaski. It was popular as a short haul transport and executive airplane. (As an example of the wide acceptance of all models of the Electra, Neville Chamberlain flew to Munich in an Electra Model 10 for his meeting with Hitler in 1938.)

Whereas the Hibbard/Johnson team designed the Model 14, the man who was project engineer for the airplane and who guided the construction of the prototype Model 14 was Don Palmer, who had left Vultee and was now working for Lockheed. The two close friends and classmates from the University of Michigan were again back together.

The Model 14, as successful as it was for Lockheed, was about to morph into another airplane that would bring the company even more fame and fortune. When 1938 rolled around, Lockheed did not have in production a military aircraft. With the clouds of war on the horizon, this was about to change. The British Purchasing Commission, the same organization that later resulted in the design and production of the P-51A (see Chapter 6) first came to the United States looking to buy a coastal patrol bomber and a trainer. Lockheed was still a relatively small company at that time, and had no expectations that the British commission would come knocking at their door. But come they did, with only a five days' notice. Lockheed reacted with the knowledge that the Model

14 had the performance and size to serve as an antisubmarine bomber, and Hibbard and staff had a full-scale wooden mock-up ready for the visit, as well as technical reports estimating the airplane's performance. The British were impressed, not only with the mock-up, but also with the enthusiasm of the small company. Lockheed was invited to come to Britain for further discussions. The Lockheed contingent was made up of only four people: Courtland Gross who was Robert Gross' younger brother and who led the team as management; Bob Proctor, the company lawyer from Boston; Carl Squier, the Lockheed vice president for sales; and Kelly Johnson. How far Johnson had come in the company – chosen as the only technical person representing Lockheed on the team! But there was no better engineer to carry Lockheed's technical expertise across the ocean; Kelly knew the airplane's design inside and out, and being the flight test engineer, he knew exactly what the airplane could so.

After presenting a 30-minute proposal at the Air Ministry, the Lockheed team was told to change completely the way that bombs and torpedoes would be installed in the airplane. In the original Lockheed design, they would be stacked in racks from floor to ceiling as standard with US Army design. The British wanted them all in the bomb bay. In addition, they wanted a gun turret for rear protection, and forward firing guns. These items may seem innocuous, but to include them in the existing design would require a major structural redesign, which in turn would affect the weight of the airplane, and which would change its performance. Undaunted, the Lockheed team did not skip a beat. They bought the essential design tools – a drawing board, T-square, triangles, and some drafting equipment – and nonstop over the next 72 hours Kelly Johnson carried out the redesign, holed up in their hotel room in Mayfair Court. He did everything necessary – fitting in the new equipment, rearranging crew positions, making weight and structural analyses, and recalculating the airplane's performance to guarantee that it would still satisfy the requirements set down by the Air Ministry. He even recalculated the contract pricing.

In so doing, Kelly Johnson was exhibiting his special talents that contribute to his being a "grand designer." The conceptual design methodology discussed in Chapter 2 and tabulated in Figure 2.7 was already incorporated in the Model 14, and the airplane was proven and flying. Johnson, however, in his 72-hour redesign process for the British, had to carry out a thought process that exercised a subset of this methodology, especially that of weight change and its effect on performance. In so doing, although still functioning at Lockheed as an engineering researcher, Johnson took yet another step closer to becoming an airplane designer. He did not design the new airplane from scratch, but its final incarnation was all due to Kelly Johnson.

The Air Ministry liked the new design. On June 23, 1938, it gave Lockheed a contract to build 200 airplanes. Labeled the Lockheed model B-14, the airplane was to become famous as the Lockheed Hudson (shown earlier in Figure 7.16). This was Lockheed's first production line military airplane, after a long and somewhat convoluted history of commercial airplanes. The Hudson was used effectively by the RAF as a coastal patrol airplane, carrying out anti-submarine and other bombing and torpedo missions. When the United States entered World War II, Hudsons were used by both the US Navy and the Air Corps. When production ended in 1943, Lockheed had produced 2,941 Hudsons.

The Hudson bomber first flew on December 10, 1938, just slightly less than six months after the Air Ministry signed the contract. At that time, the work force at Lockheed had greatly increased to 7,000. This was not, however, by any means just due to orders for the Hudson. Lockheed also had another military airplane earmarked for the production floor, the soon to become famous P-38 Lightning fighter, an airplane that had the marks of Johnson's design genius all over it.

In 1936, the US Army Air Corps issued a design competition for a twin-engine high altitude interceptor. Kelly Johnson immediately played the role of conceptual airplane designer. After considering a number of possible configurations, Johnson decided on a somewhat unorthodox design – a twin-boom configuration with an engine in each boom with the booms extending all the way back to the tail, sprouting a vertical tail from each boom, and with the pilot situated in a central nacelle mounted on the wing between the booms. In this fashion, the P-38 Lightning was born; its unconventional configuration is seen in the three-view shown in Figure 7.17. About this configuration, Kelly Johnson wrote much later:[38]

> It was considered a radically different design – even funny looking some said. It wasn't to me. There was a reason for everything that went into it, a logical evolution. The shape took care of itself. *In design, you are forced to develop unusual solutions to unusual problems.*

I have added the emphasis in the last sentence, because it might be considered Kelly Johnson's mantra as a "grand designer;" it helps to

[38] Johnson, Kelly: *More Than My Share of It All*, p. 71.

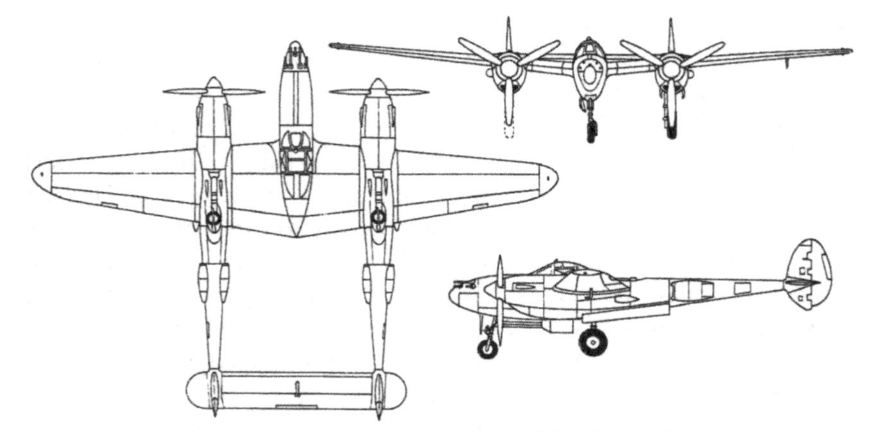

FIGURE 7.17. Three-view of the Lockheed P-38 fighter.

explain his design philosophy and the novel aircraft he designed later during the cold war. The P-38 was the first reflection of his totally different, innovative design philosophy. The total conceptual design of the P-38 was a product of the Hall Hibbard/Kelly Johnson team, working together as they did earlier on the Model 14 design. But Johnson was clearly taking an expanding role as an airplane designer. The design used Fowler flaps, so effectively employed on the Model 14that they became the first high-lift maneuvering flaps used on any fighter aircraft. The two propellers were counter-rotating – rotating in opposite directions to eliminate the rolling torque associated with propeller-driven airplanes. The airplane was designed with flush rivets, the ploy so effectively used by R. J. Mitchell for his Supermarine racers and the Spitfire a few years earlier. The power required to meet the specifications for the new Army fighter dictated the use of two engines. During the conceptual design process, different configurations were examined, including one with an engine at the front of the fuselage "pulling" the airplane and the other at the rear "pushing" the airplane, called a tractor–pusher configuration (a push–pull arrangement). A more conventional arrangement called for both engines mounted in the usual style in the wings. However, the liquid-cooled Allison engine required a long Prestone radiator and a turbocharger. This, along with making room for the landing gear to retract into the engine nacelle, made the nacelle so long that it reached far back behind the wing, almost to the tail. In this case, sensible design dictated that the nacelles become twin-booms with the tail attached at the end. Once again we see the use of new but somewhat proven technology by airplane designers – no real change in the intellectual methodology of conceptual airplane design, but rather the boldness to employ new technology.

The Air Corp was pleased with the Lockheed design, and a contract for one XP-38 prototype was issued on June 23, 1937. Detailed design was entrusted to project engineer James Gerschler, and the prototype was ready in December. The first flight of the XP-38 took place on January 27, 1939, with Army Lieutenant Benjamin S. Kelsey at the controls. Kelsey was the project officer for the Army at Wright Field, and the very fact that he flew the XP-38 on its first flight, rather than a Lockheed test pilot, was indicative of his excitement with the new fighter design. The first flight was only thirty-four minutes, and even though the airplane suffered from a violent vibration of the flaps and failure of three out of the four flap support rods, problems readily solved before the next flight, the XP-38 was considered a success. Kelsey's excitement about the XP-38 continued; just fifteen days after the first flight, Kelsey attempted an unprecedented cross-country flight from March Field in California to Mitchell Field in New York. He almost made it. After an elapsed time of seven hours and forty-three minutes – just fifteen minutes more than the transcontinental record set by Howard Hughes in his H-1 racer two years earlier – with Mitchell Field in sight the XP-38 lost power and crashed on a golf course just 2,000 ft. away from the runway. The prototype was a total loss, but Kelsey escaped unhurt. Moreover, his enthusiasm for the P-38 continued undaunted; on April 27, 1939, he awarded Lockheed an Army contract for $2 million for thirteen YP-38s. The P-38 program was underway, to end only after nearly 10,000 were produced (inclusive of all models) during World War II.

Beyond its technical firsts, the P-38 was the right airplane at the right time for service during World War II. It was:[39]

1. Lockheed's first military aircraft to go into series production. (The Hudson was the second.)
2. The first twin-engine fighter to go into service with the US Army Air Corps.
3. The first fighter with a top airspeed of more than 400 mph. (The XP-38 had a maximum speed of 413 mph, and the last model to go into production, the P-38L, could reach 414 mph.)

[39] Angelucci, Enzo, *The American Fighter*, Onion Books, New York, 1985, pp. 127–128.

4. The first fighter to be flown from the United States to Europe.
5. The first American airplane to shoot down a German aircraft.
6. The first American fighter to carry out an escort mission to Berlin.
7. The first American airplane to land in Japan after the surrender.
8. The heaviest American fighter of the War. (The gross weight of the P-38L was 17,500 lbs.)
9. The only American fighter which was in production on the first and last days of the War.

The P-38 made Lockheed a household name during the war. Its designers, Hall Hibbard and Kelly Johnson, were not so. With the exception of R. J. Mitchell in England, whose national and international fame started with the Schneider Trophy racers and was solidified with the Spitfire, none of the "grand designers" in this book were household names. For the most part, they stayed behind the scenes, although they were all well known by the professional aeronautical engineering world in their time. This held true for the young Kelly Johnson, who by 1941 had published six papers in the *Journal of the Aeronautical Sciences*, and had been elected to Associate Fellow of the Institute of the Aeronautical Sciences. (Hall Hibbard was an IAS Fellow by that time.) Johnson's position as chief research engineer at Lockheed provided him an early professional status, but still not a household name.

Because of the P-38's high speed, it encountered another first – one not so glorious – the first airplane to suffer from massive aerodynamic compressibility problems. Beginning as early as 1918, the NACA carried out a program of definitive research on the high-speed aerodynamic characteristics of airfoils, attempting to explain why, beyond a certain freestream Mach number (the Critical Mach number), a given airfoil would exhibit a drastic increase in drag and loss of lift.[40] By 1935, the NACA aeronautical engineer John Stack and his colleagues at the Langley Memorial Aeronautical Laboratory had uncovered the basic physical phenomena causing these adverse compressibility effects, and his supervisor Eastman Jacobs gave a seminal paper on this research at the 1935 Volta Conference in Rome; the topic for the conference was "High Velocities in Aviation." In this paper, for the first time, the full explanation of the physical aspects of the flow field that caused the adverse compressibility

[40] Anderson, John D., Jr., *A History of Aerodynamics*, Cambridge University Press. 1997, pp. 382–403. This section describes in depth the NACA research on airfoil compressibility effects in the 1920s and 30s.

effects on airfoil performance were revealed to the engineering community.[41] The culprit was shock waves that formed in a local pocket of supersonic flow developing over the airfoil when the Critical Mach number was exceeded. As proof, Jacobs showed the first pictures ever taken of this shock wave on an airfoil, and the consequent flow separation downstream of the shock. At this time, the NACA was the world leader in understanding the physical nature of the compressibility problem. During his conceptual design of the P-38, Kelly Johnson was fully aware of the NACA research and the problem posed by compressibility for high-speed airplanes. He used the knowledge and data generated by the NACA during his conceptual design of the P-38; indeed, in the resulting proposal to the Air Corp, Johnson wrote a discourse on the probable effects of compressibility on the P-38, referencing two NACA documents:

Stack, John and von Doenhoff, A. E., "Tests of 16 Related Airfoils at High Speeds," NACA Technical Report No. 492, 1934.

Stack, John, "The Compressibility Burble," NACA Technical Note 543.

From this research, Johnson knew that the adverse compressibility effects could be delayed to a higher Mach number by using a thinner airfoil. The NACA 23018 airfoil had been successfully used on the Super Electra Model 14, but this 18 percent thick airfoil was too thick for high-speed use. So the P-38 was designed with a thinner 16 percent NACA 23016 airfoil in the center section of the wing between the two booms, and a thinner yet 12 percent NACA 4412 airfoil for the wing shape outboard of the booms. With this thinner wing, Johnson hoped that compressibility problems would be minimized for the P-38. Nevertheless, when production models of the P-38 began to encounter massive and sometimes fatal compressibility problems in high-speed power dives, it was disappointing to, and somewhat unexpected by, Johnson and the Lockheed engineering team.

The particular problem that reared its head on the P-38 and that caused so much trouble, however, was not the usual compressibility effect causing large drag-divergence beyond the critical Mach number.[42] This large increase in airplane drag did in fact occur, but was accounted for in the

[41] Jacobs, Eastman, "Methods Employed in America for the Experimental; Investigation of Aerodynamic Phenomena at High Speeds," NACA miscellaneous paper 42, 1936. This is the paper that Jacobs delivered to the Volta Conference in Rome in 1935.

[42] Foss, R. L., "From Propellers to Jets in Fighter Aircraft Design," AIAA Paper 78-3005, in *Diamond Jubilee of Powered Flight: The Evolution of Aircraft Design*, Jay D. Pinson, editor, Dayton-Cincinnati Section, AIAA, Dec. 1978, pp. 51–64.

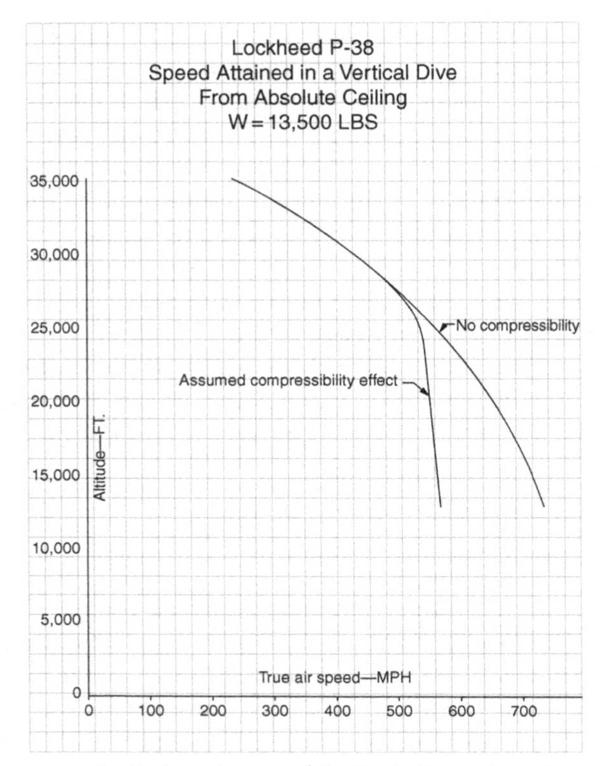

FIGURE 7.18. Early estimates of the P-38 dive trajectories, 1940.

conceptual design of the P-38. For example, the preliminary graph (Figure 7.18) prepared by Phil Coleman, the aerodynamicist working with Johnson, gives the predicted speed versus altitude dive trajectories for the P-38, comparing the analytical results for no compressibility effect with those for the assumed compressibility effects.[43] In order to verify these results, Lockheed began a flight-test dive program in the summer of 1941, and this is when things began to unravel. During some of these tests, the pilots were unable to pull out of the dive, no matter how hard they pulled back on the control stick. One of Lockheed's most valuable test pilots, Ralph Virdlen, crashed to his death during one of these tests in November. The problem being encountered by Virden and other P-38 pilots was this: beyond a certain speed in a dive, the elevator controls suddenly felt as if

[43] Ibid, p. 57.

they were locked, and to make things worse, the tail suddenly produced more lift, putting the P-38 into an even steeper dive. This became known as the "tuck-under problem," and at the time, nobody, including Kelly Johnson, knew how to explain it, much less how to fix it.

Lockheed consulted various aerodynamicists, including von Karman at Caltech, but to no avail. The new Lockheed wind tunnel could not get to the high speeds required to study the problem. The NACA 8-foot High Speed Wind Tunnel (HST) at Langley, however, could produce an air stream of 575 mph (Mach 0.75). In December 1941, just a few weeks after Ralph Virden was killed, a one-sixth scale model of the P-38 was mounted in the NACA HST and extensive testing began.[44] Under the direction of John Stack, it was not long before the NACA engineers observed shock waves forming on the upper surface of the P-38 wing, with the expected consequent loss of lift and increase in drag. More importantly, however, with this accumulated experience in compressibility effects, John Stack was the only one to properly diagnose the problem: when the shock waves formed on the P-38 wing, causing a loss of lift, the downwash angle of the flow trailing behind the wing decreased. That, in turn, increased the effective angle of attack at which the flow encountered the horizontal tail, increasing the lift on the tail and pitching the P-38 to a progressively steeper dive, totally beyond the control of the pilot.[45] Stack's solution was to place a special small wedge-shaped flap under the wing, to be deployed only when those compressibility effects were encountered. The flap, called a dive-recovery flap, was not a conventional dive flap intended to reduce the airplane's speed; rather, it was used to maintain lift in the face of the compressibility effects, thus reducing or eliminating the change in the downwash angle and therefore allowing the horizontal tail to function properly.

This is a graphic example of the vital importance of the NACA compressibility research as real airplanes began to sneak up on Mach 1. Indeed, the P-38 was the first airplane to penetrate well beyond its critical Mach number in a powered terminal dive, as seen in Figure 7.19.

The specific cause of the tuck-under problem, however, was not fully understood by Johnson and his staff. Indeed, as late as 1985, Kelly wrote:[46]

> We decided that if we could not solve compressibility, we would discover a way to slow the airplane to a speed where the effect was no longer a factor. The answer

[44] Hansen, James R., *Engineer in Charge: A History of the Langley, Aeronautical Laboratory* 1917–1958. NASA SP-4305, 1987, pp. 250–251.

[45] Anderson, John D., Jr., p. 406.

[46] Johnson, Kelly: *More Than My Share of It All*, p. 76.

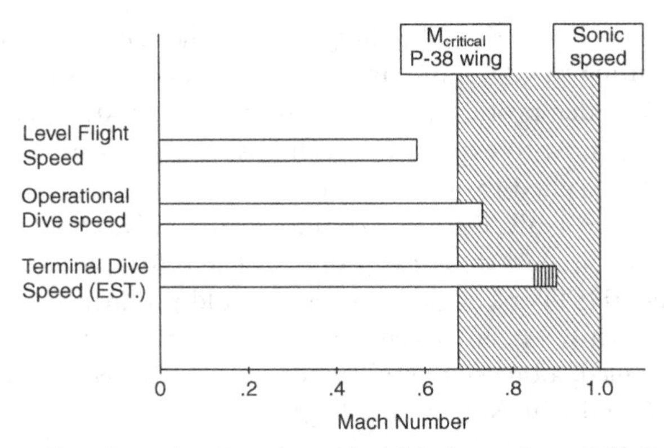

FIGURE 7.19. Bar chart showing the critical Mach number of the P-38. (R.L. Foss.)

was external dive flaps, or brakes. Put in the right place, they would cause the nose to come up out of a dive and stop buffeting.

In reality, the function of the dive-recovery flap on the P-38 was not to slow the airplane to a speed below the critical Mach number, but rather to increase the lift and hence the downwash angle while still flying beyond the critical Mach number. John Stack and his colleagues at the NACA fully understood the aerodynamic effect of the dive recovery flap, and they were responsible for this aerodynamic palliative that mitigated the P-38 tuck-under problem.

For the rest of his life, Johnson never saw it that way, and here we have an early insight to the growing self-assured and even self-centered personality of this "grand designer." Johnson was the official chief research engineer of Lockheed at this time. He was supposed to know and understand the answers, especially to aerodynamic problems. He co-opted the dive-recovery flap as the *Lockheed* dive recovery flap, and gave the NACA little credit. Indeed, writing in 1985, well after his retirement, Johnson derides the NACA:[47]

The agency charged with assisting, coordinating, and instituting this nation's aeronautical development did not want to acknowledge the work as industry-initiated. Later, NACA did do some testing on its own but had contributed nothing to solving the problem of compressibility on the P-38 except allowing the use of its wind tunnel. And this only under orders from the Army Air Corps. The successor agency, National Aeronautics and Space Administration (NASA),

[47] Ibid, p. 78.

> by contrast, has been very aggressive and eager to assist and work with industry. I am happy to report that I enjoy excellent relations with NASA.

Here, Johnson is showing his miffed feelings about being originally stood up by the NACA when he first requested wind tunnel time for testing the P-38 in their high-speed wind tunnel. Being by nature a conservative agency now entrusted with a massive wartime wind tunnel program for all US aircraft, the NACA was rightfully concerned about the P-38 model coming loose and damaging the high-speed tunnel. In a few preliminary tests, the P-38 model did indeed violently thrash around. So Johnson found the ear of the head of the Army Air Forces, General H. H.: "Hap" Arnold, who guaranteed the NACA that the Army would fix any damage to the tunnel that might occur. The NACA proceeded with the tests, and John Stack and his colleagues threw themselves into solving the tuck-under problem, ultimately coming up with the dive-recovery flap.

Thirty-five years later, Richard L. Foss, of the Lockheed-California Company in Burbank, in a survey paper on Lockheed's fighter aircraft design, takes a middle-of-the-road position:[48]

> Further P-38 wind-tunnel testing finally resolved the means for dive recovery. A dive flap was perfected, mounted on the lower surface of the main beam of the wing. When deployed, it generated a positive pressure field that created a lift and nose-up moment at high Mach numbers, and assisted in dive recovery. The device was installed on production P-38 aircraft in 1944. It did not break the descent speed – its purpose was to provide a means for recovery, and actuation was almost instant, being electrically activated. Because of their effectiveness and simplicity of operation they permitted the use of an expanded flight envelope, and increased the dive capability of the P-38 airplane.

Foss gives no explicit credit to either Kelly Johnson or the NACA for the solution to the problem, implying that, working jointly, the solution was found.

By this time, however, Kelly Johnson was preoccupied with another high-speed airplane design project, one of far more reaching significance – the design of America's first operational jet fighter, the P-80.

THE GOOD LIFE

Kelly Johnson's meteoric professional success at Lockheed during the 1930s was a source of great personal pride; he was smart, highly

[48] Foss, R. L., pp. 58–59.

knowledgeable, and exceptionally motivated. He knew it, and so did the people at Lockheed. As chief research engineer, he was given an exceptional degree of professional freedom to solve mathematical and aerodynamic problems that arose across the board at Lockheed, and he began to play a stronger role in the conceptual design of new airplanes, working with his boss, chief engineer Hall Hibbard. The intellectual methodology used by Johnson and Hibbard for the conceptual design of new airplanes was the same as first outlined by Barnwell in 1916, but as in the case with grand designers, both men were attuned to the latest developments in technology. Indeed, Johnson's close association with the relatively new Institute of Aeronautical Sciences (IAS), the purpose of which was to disseminate the latest aeronautical research results to the technical community, kept him well informed. Johnson was made an Associate Fellow of the AIAA in 1939; Hall Hibbard, already an Associate Fellow, was elected to Fellow of the IAS, also in 1939. So these gentlemen were in the mainstream of aeronautical technology, and Kelly Johnson's six papers in the *Journal of Aeronautical Sciences* before 1941 further connected him with the research community. Such a close connection made him stand out from the community of airplane designers. Our previous "grand designers," with the exception of Frank Barnwell, published very little in the open literature; it simply was not part of their culture. Kelly Johnson was setting a new cultural standard for airplane designers.

Johnson, however, was still a designer at heart. His first major professional award came in 1937. He was honored by the IAS with the prestigious Lawrence Sperry Award for "important improvements of aeronautical design of high speed commercial aircraft." This stemmed from his work with the Fowler Flap for the Model 14 Super Electra; with Johnson's modifications to this flap, he was justified in calling it the "Lockheed-Fowler Flap."

Johnson's "good life" in the 1930s also involved close contacts and associations with some of the most famous aviators of that period. Amelia Earhart flew the Lockheed Vega, setting numerous distance and endurance records between 1928 and 1935. When Johnson first met her, Earhart had a Lockheed Model 10E, the original Electra but with more powerful engines, with which Johnson carried out calculations showing Amelia how to obtain the maximum range from the airplane for her around the world flight. He calculated fuel loads, distance, and gross weight for each leg of the flight. He wrote her a long letter recommending the appropriate flap settings and takeoff procedure. Earhart's around-the-world flight ended on July 3, 1937 when the last faint radio messages were heard from her and her navigator, Captain Fred Noonan, as they desperately searched for their landing and refueling site on tiny Howland Island in the South Pacific. To this day stories and fables persist about what happened to Earhart. Johnson's matter-of-fact opinion was:[49] "The two had a rubber dinghy with them, and if undamaged, the plane would have floated with its gas tanks emptied. I am convinced that they attempted to ditch the airplane and didn't get away with it."

Because they frequented the Lockheed factory, Johnson also got to know other famous aviators. Charles Kingsford-Smith, the Australian pilot who set several distance records over the Pacific, became a good friend. Colonel Roscoe Turner, winner of the Bendix and National Air Races, Anne and Charles Lindbergh, Laura Ingalls, and Ruther Nichols all flew Lockheed aircraft and crossed paths with Johnson. What a heady experience for a budding airplane designer, not yet 30 years old!

During this period, Johnson's desire for school and more education led him to take graduate courses at the California Institute of Technology. "It is important for an engineer to keep up with advancing technology," he wrote. "Studying, fortunately, still held for me the same fascination that it had when I discovered the Carnegie library in Ishpeming."[50] He attended Wednesday afternoon seminars, listening to prominent engineers and scientists. Caltech was rapidly becoming the second most prestigious technical university after MIT. von Karman had arrived from the University of Aachen, and along with Clark Millikan, head of the Department of Aeronautical Engineering, was teaching a series of very modern courses at Caltech, including a pioneering course on high speed aerodynamics and compressible flow – exactly what Johnson would need for the future. Kelly took a number of classes, over time completing all the course work requirements for a Ph.D. The foreign language requirement in technical German, however, would require much more study and preparation time than Johnson's increasingly busy schedule at Lockheed would allow, and he stopped right there. In terms of formal graduate education, Johnson had gone further than any of the "grand designers" that we have studied in this book. Indeed, to my knowledge, no known airplane designers until recent time had earned a Ph.D.; they simply did not need it for their time and place. Even a master's degree, such as held by Art Raymond, was unusual among airplane designers. So Kelly Johnson's completed coursework for the Ph.D. at Caltech put him in a class by himself, and was to

[49] Johnson, Kelly: *More Than My Share of It All*, p. 46. [50] Ibid, p. 68.

serve him well later in the jet age. The very fact that he simply craved studying and wanted to take graduate classes for the pure sake of learning is reflective of the unique mental personality of this airplane designer.

Johnson's personal life also matured. When he first joined Lockheed in 1933, he met and dated Althea Louise Young, Lockheed's postmaster and assistant treasurer. Kelly fell in love with this tall, attractive young woman, and she was attracted to this young, "brainy" (as he was described to her by her female colleagues) up-and-coming engineer. At that time, she made more money than Johnson. Only when economic parity was reached did Kelly consider marriage to be a viable option. They were married in 1937 and rented a house in the foothills on Country Club Drive in Burbank. Three years later they were able to buy a lot on Oak View Drive high in the hills of Encino. From his early days with his father in Michigan, Kelly knew construction. Moreover, Althea and Kelly were both self-sufficient individuals. They built their house on this lot mostly by themselves, even including a swimming pool and tennis court to cater to their athletic nature. They were in good company; Clark Gable owned the property on the hill above them, and Alice Faye and Phil Harris lived nearby. The mayor of Encino at that time was Al Jolson. It is no surprise, therefore, that the social circle of Althea and Kelly reached well beyond that of aeronautical engineers and an airplane company to include the other major business of the Los Angeles area – show business.

During World War II, the Johnson's bought the Lindero Ranch, consisting of 226 acres of rolling country. On this property they built a second house – a ranch house on a mountain top with a view of the Pacific Ocean to the southwest and a range of mountains to the northeast. They planted and harvested 110 acres of hay on their ranch to help feed a growing heard of Hereford cows, and to stable a number of horses; both Kelly and Althea shared intense interest in horseback riding.

The Johnsons never had any children but they shared a happy and supportive marriage until Althea died of cancer in December 1969. Althea wished to be cremated. Kelly piloted a small airplane over the Santa Barbara Bay, and scattered her ashes as she had wished.

KELLY JOHNSON AND THE SECOND DESIGN REVOLUTION

The Lockheed Aircraft Corporation entered the jet age unilaterally, due to the inventive nature of one of their propulsion engineers. Nate Price worked with turbo superchargers and had experience with industrial turbine power plants. In 1940, with internal Lockheed company support,

FIGURE 7.20. Hall Hibbard holding a model of the Lockheed L-133 jet fighter design, 1942.

he designed a gas turbine jet engine with a high compression ratio, twin-spools, and an afterburner, prescient for its time. Nobody in the United States at that time knew about Von Ohain's jet engine in Germany, nor Whittle's engine in Britain. Price's design was unique to him and to Lockheed, and it promised an extraordinary 5,000 lbs. of thrust. Johnson credited Nate Price as "a designer of great vision and knowledge of thermodynamics, materials, and mechanical design."[51]

Lockheed undertook the preliminary design of an airplane to use this jet engine. Led by Johnson, the design team conceived a stunning looking canard configuration (Figure 7.20), made from stainless steel, and theoretically capable of 625 mph at 50,000 feet powered by two of Price's axial flow jet engines. The company's designation for this design was the L-133, and in March 1942, Kelly Johnson marched into Wright Field with an armful of design and technical data for the L-133. The Army, however, felt that the project was much too risky, and instead urged Lockheed to work as hard as possible on solving the compressibility problem with the P-38 and concentrate on other immediate wartime projects. However, they did give Lockheed a small contract to develop Price's jet engine, then labeled the XJ37, but it never got beyond the development stage.[52]

[51] Ibid, p. 95.

[52] Young, James o., *Lighting the Flame: The Turbojet Revolution Comes to America*, Air Force Flight Test Center History Office, Edwards AFB, California, 2002, p. 11.

Perhaps another reason for Wright Field's dismissal of the Lockheed proposal was that, completely unknown to Johnson and Lockheed, in September 1941, the Army had given a contract to General Electric to build a copy of Whittle's jet engine, and to Bell Aircraft to build an airplane to use it, namely, the Bell XP-59. The most important outcome of Johnson's visit to Wright Field was that he had made an impression with his extensive knowledge about jet-propelled airplanes – an impression that was to serve him very well the following year.

In 1940, both Hall Hibbard and Kelly Johnson made a decision internally within the company that the propeller, which encountered severe compressibility effects at the tip, was on its way out for use in a high performance fighter airplane. The compressibility problems encountered with the P-38 brought home this limitation. But for the time being, they kept these feelings to themselves. Then, in 1941 Johnson published another paper in the *Journal of the Aeronautical Sciences*; entitled "The Design of High-Speed Military Airplanes." It was completely different in nature from his previous papers in the journal.[53] Previously his papers were focused on detailed technical aspects based on his engineering and flight test work at Lockheed; this one, however, was part summary of the state of the art of fighter airplane design, part description of the technical problems in airplane design at the time, and most importantly, part philosophizing and looking into the future of fighter design. At the time he wrote the paper, his mind was deeply involved with the L-133 jet fighter design and he was convinced that jet propulsion was the wave of the future. He could not, however, write anything that might reflect on this secret project. With the benefit of hindsight and knowing what was in his mind, a careful reading of his paper unearths some subtle references to what Johnson was really thinking, even though on the surface it might seem that he was going down the same well-traveled road that led to the inherent limitations of the propeller-driven airplane.

At the beginning of this paper, Johnson praises some of the detailed design developments of both British and German aircraft brought on by experience already gained at the beginning of World War II – for example, the British success in using the eight-gun fighter (the Spitfire and Hurricane), the incorporation of heavy armor protection for the pilot, self-sealing fuel tanks, and the successful use of the dive bomber.

[53] Johnson, Clarence L., "The Design of High-Speed Military Airplanes," *Journal of the Aeronautical Sciences*, Vol. 8, No. 12, Oct. 1941, pp. 467–474.

Then he reflects on the use of new technology in airplane design under the exigencies of wartime, warning that:

> The element of time is of paramount important under wartime conditions because the manufacturer, or, more properly, the nation cannot afford to gamble too far on new and untried design features for fear that trouble experienced in their development may make his airplane obsolete and waste his production facilities at a time when the greatest output is required. It is better to have 1000 four-hundred m.p.h. airplanes in service than to have the finest four-hundred and fifty m.p.h. design in the world on paper struggling with problems of design and production.

This is a reason why Johnson should have understood why the Army officials at Wright Field half a year later turned down his proposal for the L-133 jet airplane design. But Johnson cushions the above statement in the same paragraph:

> Only after sound fundamental research and the highest type of engineering can the manufacturer afford to use some of the newer developments now on the horizon of progress.

Johnson knew that on Lockheed's horizon was the jet airplane and that the company's "sound fundamental research and the highest type of engineering" was being directed to efforts that would lead beyond the L-133, ultimately to the P-80 Shooting Star.

Johnson emphasized the importance of drag reduction, writing: "The relation of airplane drag and thrust horsepower for airplanes in the 400 m.p.h. class requires that extremely careful attention be given to the reduction of air resistance if the airplane size and power are to remain within the bounds of practicality." He goes on to say: "The designer must continually attempt to make the smallest possible airplane for a given power and be sure that the drag of every element is as low as can be obtained without sacrificing stability, control or maximum lift ... The drag of the various elements of the airplane has been reduced greatly from values in existence several years ago."

At this point, Kelly Johnson was aware of one of the NACA's most important and useful activities, the drag-cleanup problem carried out in the Langley Full Scale Tunnel. Beginning in 1938, the NACA put whole fighter airplanes in the tunnel, starting with a completely faired condition with all exterior devices and protuberances removed, and then one by one restoring the airplane to its original configuration, measuring the drag at each step and thus isolating the drag of each element. This program remained in effect throughout most of World War II and provided airplane designers with the most detailed knowledge of what elements were causing what drag, and allowing the designers to clean up their airplanes, resulting in substantial increases in the airplanes' performance.[54] Virtually all the important aircraft companies took advantage of the NACA's drag cleanup program, and in his 1941 paper, I can only surmise that by not mentioning it, especially in the context of his discussion of the importance to the designer of being "sure that the drag of every element is as low as can be obtained", Johnson is either subconsciously or consciously ignoring this important NACA contribution.

In his same discussion on drag, Johnson gives a nod to another important research program, but again not mentioning the NACA nor giving it any credit for its work – the laminar flow airfoil. He wrote:[55]

> There is still a possibility of substantial reduction in the skin frictional resistance, roughness and cooling drag. By using airfoils with shapes which extend the laminar flow area the drag can be reduced considerably, but these wing sections must be considered from the point of view of manufacturing difficulties, maximum lift and the effect of roughness which is inevitably encountered in service when flying off rocky fields or through hail storms.

He is somewhat optimistic, however, about the future use of laminar flow airfoils, noting that: "There is an undeveloped field requiring much additional research on this problem which seems to show much promise."

Eastman Jacobs, one of the NACA's leading aerodynamicists, developed the laminar flow airfoil shapes in the late 1930s. In fact, Jacobs was responsible in the early 1930s for the development of the NACA four-digit airfoil family, and in the middle 1930s for the five-digit airfoil shapes – both of which Kelly Johnson used for the Super Electra and on the P-38. (The P-38 had an NACA 23016 airfoil inboard of the twin-booms and an NACA 4412 airfoil outboard.) In 1938 Jacobs designed and tested a new family of airfoil shapes intended to encourage laminar flow over a substantial length of the airfoil. The results were dramatic; laminar flow was achieved in some cases over 60 percent of the airfoil, with a marked decrease in the skin-friction drag coefficient. With war clouds gathering on the horizon, the NACA classified these results. Nevertheless, in the 1939 annual report of the NACA, a cryptic statement was made in conjunction with experiments made in the new Langley

[54] Anderson, *The Airplane: A History of Its Technology*, pp. 213–215.
[55] Johnson, Clarence, "The Design of High-Speed Military Airplanes," p. 469.

low-turbulence wind tunnel: "These preliminary investigations were started by the development of new airfoil forms that, when tested in the new equipment, immediately gave drag coefficients of one-third to one-half the values obtained for conventional sections." Even though the new laminar-flow airfoil data were classified, we have discussed in Chapter 6 how the NACA made their existence known to select people in the aircraft industry, and how Edgar Schmued, in 1940, adopted its use in the design of the P-51 Mustang. So when Kelly Johnson wrote in his 1941 paper that "There is an undeveloped field requiring much additional research on this problem which seems to show much promise," the P-51 was already flying with a laminar flow wing.

We have emphasized that an important characteristic of a "grand designer" is the use of new but relatively proven technology in the design. A careful study of Johnson's 1941 paper shows an author who clearly appreciates new technology, but is conservative enough to wait for it to be reasonably proven before using it in the design process. Moreover, he is clearly hampered by the need for secrecy. For example, he wrote this paper at the same time that he was feverishly working on the design of a jet-propelled airplane, the L-133 project, and when he and Hubbard agreed that the propeller was on the way out. Yet, a substantial portion of Johnson's paper was on propellers, and in particular on the adverse effects caused by compressibility on propeller efficiency. He points out that at the high speeds being achieved by fighter airplanes, not only the tips but also the hub is where shock waves occur, and that special propeller hub designs had been developed to avoid this. He predicted a trend towards wider propeller blades and the use of more blades, noting that the four-bladed propeller was coming into use. He wraps up this particular discussion with the statement that "considerable study must be directed toward obtaining propeller sections allowing higher critical compressibility speeds." His sharp edge is evident in the 1941 paper when he goes on to write:

> Research on propellers operating at very high forward speeds is entirely lacking in spite of the continual requests on the part of industry for such information from the various Government Agencies equipped to obtain this vital information.

This appears to be a slap at the NACA, but now also at the Air Force at Wright Field. By 1941, Kelly Johnson had become a recognized and respected figure in professional circles, and he could get away with such statements, mainly because he genuinely felt them.

Towards the end of his paper, he gives a sketch of three future types of fighter configurations (Figure 7.21). They are all pusher configurations,

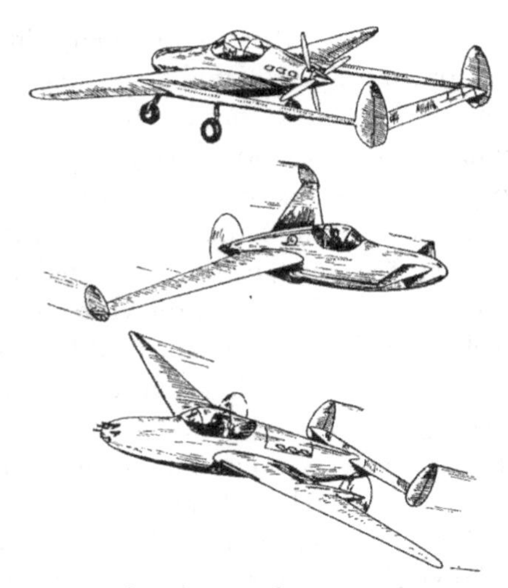

FIGURE 7.21. Three types of pusher configurations as envisioned by Johnson in 1941.

about which he writes: "In spite of the fact that the pusher propeller has many problems, it seems fairly certain on the basis of present data that it should be used." This seems to be an almost disingenuous statement; just compare the L-133 (Figure 7.20) that Johnson was really working on with the pusher configurations in Figure 7.21. No successful propeller-driven fighter used or developed during World War II was a pusher. On the eve of the United States' entry to World War II, was Kelly Johnson deliberately trying to mislead the readers, both foreign and domestic? Possibly. He was by then one of the most intelligent and knowledgeable airplane designers in the world. Is there any other reason why he would gamble his growing reputation except to mislead the opposition? This remains a curious open question about Kelly Johnson. (In a recent opinion editorial in Flying Mag.Com, January 2011, pp. 79–80, the noted and respected aviation writer Peter Garrison makes the same observation. Garrison says, "The whole paper, in fact, can perhaps most charitably be interpreted as a recital of common knowledge seeded here and there with 'volitional errors,' – or deliberate disinformation.")

What is not an open question, however, is that Johnson gives a hint of his real thinking when, buried in a discussion of turbo superchargers, he notes that a small percentage of the thrust ordinarily can be obtained from the jet exhaust of the turbo supercharger. Then, in a prediction about which he knew a lot more than he could divulge, he wrote: "As speeds increase and propulsive efficiency decreases due to compressibility effects, it is not out of the realm of possibility to consider engines which furnish their power for take-off and climb to a propeller which can be feathered for level of flight, and, in the latter condition, *the total engine power output being used by highly developed jets for propulsion.*" We know now that as Johnson wrote this statement about jet propulsion, he already had one foot, if not both feet, firmly planted in the new emerging jet age.

THE P-80 SHOOTING STAR, AND THE ORIGIN OF THE SKUNK WORKS

Kelly Johnson quickly adjusted to the Air Force's rejection of his design for the L-133 jet propelled fighter. Taking his own advice as given in his 1941 paper that "the nation cannot afford to gamble too far on new and untried design features ..." he backed away from the advanced configuration of the L-133 and began to design a jet fighter with a more conventional airframe. The result was the P-80 Shooting Star. A three-view of the P-80A, the first mass-produced model of the Shooting Star, is shown in Figure 7.22; comparing this with the Model L-133 in Figure 7.20, we clearly see that Johnson returned to a more conventional approach.

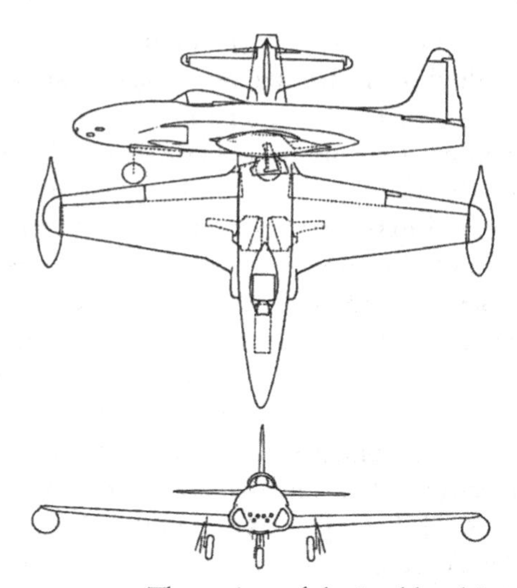

FIGURE 7.22. Three-view of the Lockheed P-80A.

During the months following the Air Force's rejection of Johnson's L-133 proposal, a lot happened to completely change the picture. In the skies over France and Germany, two new German fighters began to appear with disconcerting regularity, both powered by jet propulsion rather than propellers. The stubby, rocket-powered Me-163 was more an oddity than a real threat to Allied bombers, but the twin-engined gas turbine powered Me-262 was in a different category. This jet fighter, the first of its kind, flew about 100 mph faster than the best British and American propeller-driven fighters and posed a serious threat to the entire Allied strategic bombing campaign. General Henry "Hap" Arnold, then Chief of the US Army Air Corp, had already witnessed the British Gloster E28/39 in flight during a visit in April 1941, and immediately recognized the great potential of such a jet-powered airplane. To his knowledge, the United States had neither a jet engine nor a jet-powered airplane anywhere in the works. (At that time Lockheed had not yet shared with anybody their preliminary work on Nate Price's jet engine design nor the L-133 airframe to use it.) (James Young, *Lighting the Flame: The Turbojet Revolution Comes to America,* Air Force Flight Test Center History Office, 2002) properly notes that Hap Arnold should not have been surprised by the jet-propelled airplane. Jet propulsion for airplanes had been discussed as a possibility by the aeronautical engineering community since the 1920s. He quotes Ezra Kotcher, then serving as the senior instructor at the Air Corps Engineering School at Wright Field, as saying "it reached the point that you couldn't throw a whiskey bottle out of a hotel window at a meeting of aeronautical engineers without hitting some fellow who had ideas on jet propulsion." Young notes that Kotcher had submitted a report to General Arnold's office (Air Corps Material Division Engineering Section Memorandum Report 50-461-351) recommending an extensive transonic research program using gas turbine or rocket-powered airplanes. Apparently, his recommendations were ignored by General Arnold's staff. Ezra Kotcher later became the driving force in the Army for the eventual design of the rocket-powered Bell X-1 in which Captain Charles "Chuck" Yeager first broke the speed of sound on October 14, 1947.) Arnold, however, moved fast. He brought back with him the blueprints for the Whittle engine and arranged to have General Electric build a copy by late summer of 1941. With major improvements, General Electric produced its I-A centrifugal flow jet engine, and the Air Force moved quickly to have the Bell Aircraft Company design an airframe to use it. The Bell XP-59, America's first jet-propelled airplane, finally flew for the first time on October 2, 1942. Its performance, however, was disappointing. Primarily due to a lower than expected thrust output from the engine and the high drag from the rather ungainly-looking airframe with a very large wing area, the XP-59 could reach a maximum speed of only 404 mph, slower than the frontline propeller-driven fighters of the day such as the P-38 Lightning and the P-51 Mustang. Looking for a better jet airplane, the Air Force, remembering Kelly Johnson's proposal for the L-133 months earlier, returned to Lockheed on May 17, 1943, with a request for a new jet airplane design.

Kelly Johnson was ready. During the interim, he had carried out the conceptual design of the XP-80. Within a week after Wright Field's request, Johnson was back in Dayton presenting his design which the Air Force quickly accepted, but with the condition that Lockheed use the British de Haviland-built Halford H.1B jet engine for the first prototype. Lockheed received the Letter of Intent, approved by Hap Arnold himself, on June 8, 1943. A more formal Letter Contract for $515,018.40 was issued on June 17, and it was approved on June 24.

At the time, Lockheed was straining at the seams, producing twenty-eight airplanes of various types per day. The subsequent dynamics within the Lockheed management led to the formation of a small group that would spawn the now famous Lockheed Skunk Works. Johnson tells about Lockheed's President Robert Gross's reaction to this contract to design and build a jet airplane:

> You brought this on yourself, Kelly. Go ahead and do it. But you've got to rake up your own production people and figure out where to put this project.[56,57]

Gross did not fully realize that this was just what Kelly Johnson wanted – a small, independent organization with its own designers, engineers, shop technicians, and mechanics with the self-contained ability to design a new airplane, build it, and roll it out of the shop, all without the hindrance and delays that go along with the much larger parent company. Johnson built a makeshift working space adjacent to the Lockheed wind tunnel using empty wooden engine boxes for the walls and a rented circus tent for a roof, and he appropriated the wind tunnel shop, buying a small local machine shop to obtain more tools. He scavenged the company to get twenty-two engineers whose work he knew and respected. These included Don Palmer, his good friend from their days at Michigan who became one of Kelly's most valuable design engineers. Johnson recruited three other

[56] Anderson, John D., *A History of Aerodynamics*, pp. 342–352.
[57] Johnson, Kelly: *More Than My Share of It All*, p. 97.

engineers who were University of Michigan graduates. One of these was Willis Hawkins, who went on to become a well-known designer in his own right. Three engineers from Caltech – Phil Colman, Irv Culver, and E. O. Richter – also joined the team. The group even had its own purchasing department. With this, Johnson had everything he needed to operate independently of the main Lockheed factory. This group was, in essence, the origin of the "Skunk Works."

That name surfaced quickly, but not from Johnson. The origin of the name "Skunk Works" is clearly explained by Ben Rich, who much later succeeded Johnson as head of the Skunk Works. The physical location of Johnson's group was next to a plastics factory that emitted a noxious stench. Rich relates:[58]

> Around the time Kelly's crew raised their circus tent, cartoonist Al Capp introduced Injun Joe and his backwoods still into his Lil Abner comic strip. Ol' Joe tossed work shoes and dead skunk into his smoldering vat to make 'kickapoo joy juice.' Capp named the outdoor still 'the skonk works.' The connection was apparent to those inside Kelly's circus tent forced to suffer the plastic factory's stink. One day one of the engineers showed up for work wearing a civil defense gas mask as a gag, and a designer named Irv Culver picked up a ringing phone and announced 'Skonk Works.' Kelly overheard him and chewed out Irv for ridicule: 'Culver, you're fired,' Kelly roared. 'Get your ass out of my tent.' Kelly fired guys all the time without meaning it. Irv Culver showed up for work the next day, and Kelly never said a word.
>
> Behind his back, all of Kelly's workers began referring to the operation as the 'skonk works,' and soon everyone at the main plant was calling it that too. When the wind was right, they could smell that 'skonk.'

In 1960, Al Capp's publisher objected to Lockheed's use of the word "Skonk" and the company quickly changed it to "Skunk Works," registering the name and logo as trademarks.

Kelly Johnson's design genius and his entire Skunk Works team thrived in this new environment. Moreover, the Air Corps was highly motivated to get a viable American jet engine fighter into the air and closely cooperated with and supported the activity of the Skunk Works. Just 141 days after the signing of the formal contact with the Air Corps, the Skunk Works had designed, built, and completed preliminary ground tests on the XP-80. The first flight was delayed, however, when during a final engine test, the intake ducts collapsed under the pressure difference

[58] Rich, Ben R., with James, Leo, *Skunk Works*, Little Brown and Company, Boston, 1994, pp. 111–112.

created by low pressure flow inside the intake and higher atmospheric pressure outside the ducts. Pieces of metal were sucked into the engine, cracking the compressor housing. The De Haviland engine was ruined, and moreover, it was the only engine that Lockheed had available at the time. Several weeks went by before a new engine arrived. Finally, on that cold January morning in 1944, as related at the beginning of this chapter, 198 days after signing the contract, the XP-80 "Lula Bell" took to the air with Lockheed test pilot Milo Burcham at the controls.

The design of the XP-80 and subsequent versions of the Lockheed P-80 by itself lifts Kelly Johnson into the panoply of the "grand designers." There was nothing new in his methodology of conceptual airplane design, but the XP-80 was a masterful composition of Johnson's use of new technology. First and foremost was the use of a jet engine instead of the conventional piston engine/propeller combination. The major impact of the jet engine during the conceptual design process was purely and simply the availability of a power plant with considerably more thrust, resulting in prediction of much higher maximum velocities. Indeed, in Appendix C, the thrust produced by one of the supercharged Allison reciprocating engine/propeller combinations used by Kelly Johnson for the P-38 is calculated and compared with the thrust produced by the Allison jet engine used by Johnson on the P-80. (This is a one-on-one comparison, one jet engine compared to one reciprocation engine.) The calculation is made for a flight velocity of 420 mph at an altitude of 20,000 feet. The calculations show that the single jet engine produces *twice as much thrust* as the single reciprocating engine/propeller combination. This drives home the advantage of a jet engine; it is clearly a *high thrust-producing power plant*. Of course, this feature was well appreciated by Kelly Johnson and others at Lockheed. He did not hesitate to wrap the P-80 design around this very new technology.

The jet engine had an impact on the external aerodynamic configuration of the P-80. Because the engine was mounted inside the fuselage behind the cockpit, air had to be fed in from the outside, through the fuselage, and into the front of the engine. Instead of taking the obvious approach of using an open inlet at the nose of the fuselage and ducting the airflow straight through to the engine, Johnson instead designed two side inlets on either side of the fuselage, slightly ahead of the wing juncture, ducting the air around both sides of the cockpit, and smoothly joining the two streams of air at the entrance to the engine compressor. The side inlets are clearly shown in the three-view of the P-80A shown in Figure 7.22. The external shape of these inlets had a serendipitous effect: they gave the P-80 a slightly higher critical Mach number.[59,60] The internal flow was a different story, however. During early flights of the P-80, a disconcerting rumble could be heard coming from the air inlets. Wind-tunnel tests showed uneven flow coming into the two ducts, causing a "snaking" action of the flow into the engine. The detailed aerodynamic flow data also indicated the root cause of the problem: flow separation occurred just inside the inlet. The Skunk Works found the solution after the additional extensive wind-tunnel tests on the inlet flow: a duct system boundary layer bleed device mounted flush with the wall just inside the inlet. This duct system flushed away all the low energy air in the boundary layer on the surface, thus inhibiting flow separation. To my knowledge, this was the first use of a duct system boundary-layer bleed device on a production airplane, and it worked well. The rumble and the snaking went away, and the quality of the flow entering the engine improved.

The inlet flow problem and its solution illustrate two growing trends in airplane design at the beginning of the jet age. First, with the boundary-layer bleed device, Kelly Johnson not only used new technology in his P-80 design, he and his Skunk Work team actually *developed* the new technology, and it was proven by them in the Lockheed wind tunnel before they incorporated it into the airplane design. Second was the important role played by wind-tunnel testing in solving this problem. Kelly Johnson was no stranger to wind-tunnel testing. By the time of the second design revolution, wind tunnel technology had improved to the point where airplane designers had enough confidence in the validity of wind tunnel data to use it in critical aspects of their designs. Edgar Schmued's adoption of a laminar low airfoil for the P-51, based on the data obtained by North American in the University of Washington's wind tunnel (Chapter 6) is a case in point.

In regard to laminar flow airfoils, we reflect again on Kelly Johnson's comment in his 1941 paper identifying such airfoils as "an undeveloped field requiring much additional research on this problem which shows much promise."[61] Just two years later, Kelly Johnson designed the P-80 with a laminar flow airfoil, the 13 percent thick NACA 65-213 airfoil. By that time, the NACA had carried out "much additional research" on the

[59] Anderson, John D., *Introduction to Flight*, McGraw-Hill, New York, 2012, p. 479.
[60] Johnson, Clarence L., "Development of the Lockheed P-80A Jet Fighter Airplane," *Journal of the Aeronautical Sciences*, Vol. 14, No. 12, December 1947, pp. 659–679.
[61] Johnson, Clarence L., "The Design of High-Speed Military Airplanes," p. 470.

laminar flow airfoil, and although the data were classified during wartime, the NACA made sure that the aircraft industry knew about it. Kelly Johnson, however, saw beyond the intended purpose of the laminar flow airfoil. He knew that surface roughness introduced by manufacturing tolerances and service in the field would most likely prompt turbulent flow over the wing and negate any hoped for advantage of laminar flow. Indeed, he raises this concern in his 1941 paper. (We have already noted in Chapter 6 that Lee Atwood raised the same concern about the laminar flow wing on the P-51.) On the other hand, Johnson recognized that the shape of the laminar flow airfoils resulted in a higher critical Mach number, a serendipitous aspect of this family of airfoil shapes that was not considered in their design. Verified by wind tunnel tests, their higher critical Mach number ultimately made the laminar flow family of NACA airfoils the choice for high-speed subsonic jet fighters in the first two decades of the second design revolution. Kelly Johnson's choice of the NACA 65-213 airfoil for the P-80 was at the leading edge of this design trend. His use of the laminar flow airfoil to obtain a higher critical Mach number for the P-80 is yet another example of a "grand designer's" incorporation of new but relatively proven technology in his airplane design.

Today, the aeronautical community thinks of Kelly Johnson in terms of his later airplane designs, the SR-71 Blackbird, the U-2 Dragon Lady, etc. However, the Lockheed P-80 Shooting Star is a jewel among all his designs. In today's modern world of airplane design, the old adage that a new airplane is based on the previous one is still sometimes the case. When Kelly Johnson designed the P-80, however, there was no "previous one." The P-80 was America's first mass-produced jet fighter. Redesignated the F-80, it was the first US jet to go into combat at the beginning of the Korean War, the first US jet to participate in jet-to-jet aerial combat, and the first jet to shoot down another jet airplane, a Russian MIG-15 supplied to the North Korean Air Force. The version produced in the largest number was the F-80C, with 799 manufactured. Its maximum speed was 594 mph at sea level, and 543 mph at 25,000 feet. This was a quantum leap above the best piston-engine fighters, such as the Lockheed P-38 with a maximum speed of 414 mph and the North American P-51D with a maximum speed of 437 mph. Kelly Johnson had put America squarely into the jet age, and he never looked back.

Johnson resumed his publishing in the *Journal of the Aeronautical Sciences* with a twenty-page paper in 1947 on the design and development of the P-80A. Once again, it was the lead paper, a testimonial to

Johnson's stature in the aeronautical engineering profession. This paper is both stunning and revealing, stunning because the paper was unclassified and approved by both Lockheed management and the Air Force, even though the P-80A had gone into service only two years prior. In today's environment, it would have been classified secret by the military, and also considered proprietary by the company. It was revealing because it is packed with details about the mechanical and aerodynamic design features of the airplane.

At the time, the jet airplane was still quite new. Johnson states his purpose in publishing the paper right at the beginning:

> As this airplane is the first American jet-propelled tactical aircraft, many lessons have been learned from its design, testing, and operation which should be of great assistance in improving future similar aircraft.[62,63]

Johnson did not hold back; his command and understanding of new technology, and the willingness to use it, are seen everywhere in this paper. For example, he gives a graph of aerodynamic drag as a function of Mach number for the P-80 – not wind-tunnel data, but actual *flight test data*, where the drag is obtained from measurements of tailpipe thrust and dive angle. This data is shown in Figure 7.23, taken directly from Johnson's paper. Clearly the drag-divergence Mach number for the P-80A is about 0.75, where the drag almost doubles as the flight Mach number increases from 0.7 to 0.8. The data is remarkable for two reasons: first, it is the *real thing*, obtained in flight with the real airplane, and second, such drag-divergence data for a given military jet fighter, especially a new design, is usually classified secret. But here we have Kelly Johnson, the academically oriented airplane designer, steeped in the open-access tradition he absorbed at the University of Michigan, genuinely wanting to share technical data and the "lessons learned" from "the first American jet-propelled tactical aircraft," and of course basking in the success of the P-80 design.

By the mid-1940s, the use of wind-tunnel data by airplane designers had become an accepted, indeed a vital, aspect of the business of designing airplanes. Johnson, no stranger to wind tunnels, noted that "the aerodynamic development of the P-80A entailed the use of all types of wind tunnels." The Lockheed wind tunnel, designed by Johnson in 1939 with

[62] Clarence L. Johnson, "Development of the Lockheed P-80A Jet Fighter Airplane," pp. 659–679.
[63] Ibid, p. 659.

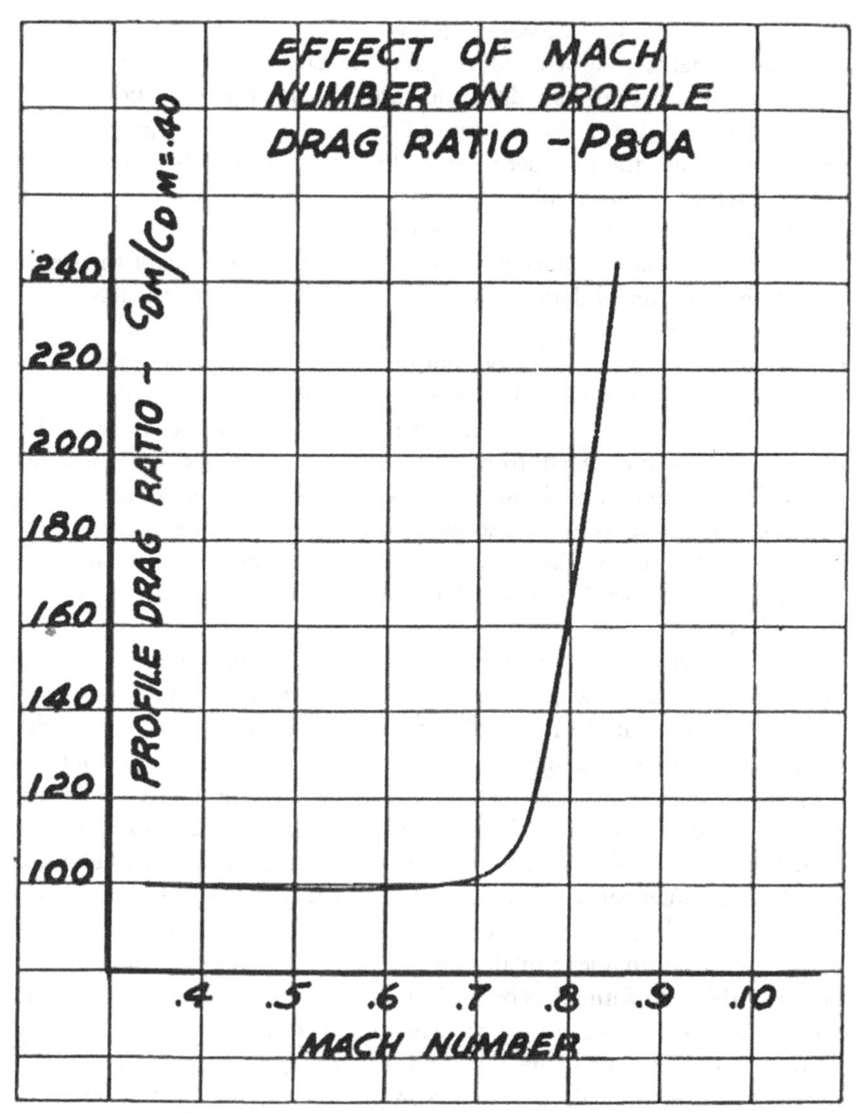

FIGURE 7.23. Flight measurements of the drag versus Mach number for the P-80A.

a flow velocity of 300 mph in its 8 by 14 foot test section, in Johnson's words had already "paid for itself" during the design of the P-38. Now with the embryonic Skunk Works staff literally sitting next to the wind tunnel, the all-out effort to design the XP-80 in 141 days, and later the

P-80A in 132 days, was expedited. Johnson orchestrated wind tunnels for the new jet fighter like a conductor for the Los Angeles Philharmonic, resulting in a variety of new design features on the airplane. One of these was the wing-tip fuel tanks, as seen in Figure 7.22. The early jet airplanes were gas guzzlers and in order to extend the range of the P-80 for its role as a fighter-interceptor, Johnson designed streamlined, teardrop-shaped external tanks to carry extra fuel. Such drop tanks were used by some propeller-driven fighters during World War II, suspended under the fuselage. With wind-tunnel data to back him up, Johnson instead placed drop tanks underneath the wing tips of the P-80. With this configuration, wind-tunnel tests showed only a 12 percent increase in the airplane profile drag compared to a 27 percent increase for a pylon-mounted tank on the fuselage. With the extra fuel provided by the tip tanks, the range of the P-80A was almost doubled, from 780 miles to 1,440 miles. Also, aerodynamically, the tip-mounted tanks increased the effective aspect ratio from 5.5 to 5.9 by acting as useful end plates on the wings. Moreover, Johnson wrote that "the tip location had minimum adverse effect on the airplane critical Mach number. The P-80A has flown to speeds over M = 0.75 with the tanks on."[64]

Johnson went against whatever little tradition there was in the middle 1940s in regard to the way that outside air from the freestream was ducted into the jet engine. In the few earlier jet airplanes, intuition had guided the designers to allow for a simple straight through duct feeding the outside ambient air directly into the inlet of the engine. This worked well for twin-engine jet fighters such as the German Me-262 (Figure 7.24) and the British Gloster Meteor (Figure 7.25). But for a single-engine jet fighter, a simple nose inlet such as used for the Republic F-84 (Figure 7.26), required that the air be ducted around the cockpit, and did not easily allow for any armament in the nose. To avoid this problem, Johnson designed the P-80 with inlets on both sides of the fuselage just ahead of the wing roots, as seen in Figure 7.22. The two air flows were then smoothly merged into the engine well behind the cockpit. Johnson showed a detailed inboard profile of the P-80A in his paper, another amazing revelation of technical data for America's first operational jet fighter. This profile, taken directly from Johnson's paper is shown in Figure 7.27.

The side inlets, however, were plagued by an aerodynamic problem anticipated by Johnson. The flow separated from the surface inside the

[64] Ibid, p. 669.

FIGURE 7.24. The Messerschmitt Me 262.
(National Air and Space Museum, SI-2005-22903.)

FIGURE 7.25. The Gloster Meteor.
(National Air and Space Museum, SI-82-3547.)

inlet, setting up an unsteady flow that created both a duct "rumble" and a directional "snaking" of the flow path. The flow entering the inlet traversed a long external path along the surface of the fuselage before reaching the inlet, resulting in a large buildup of the boundary layer adjacent to the surface. Friction robbed the boundary layer of energy, and hence this low energy flow readily separated from the surface inside the inlet. Johnson's fix for the problem was a boundary-layer bleed device installed on the fuselage wall just inside the lip of the inlet that ingested the low energy boundary layer but allowed the higher energy flow above the boundary layer to enter the inlet undisturbed. This fixed the duct "rumble" and eliminated the directional "snaking." Today, such boundary-layer bleed devices are

FIGURE 7.26. The Republic F-84E Thunderjet.
(National Air and Space Museum, NAM-A-4113-F.)

FUSELAGE CONTENTS

1. Adjustable Light
2. Oxygen Cylinder
3. Ammunition Box
4. Armament Junction Box
5. Command Radio
6. Instrument Panel
7. Bullet-Proof Windshield Panel
8. Gun Sight
9. Seat
10. Fuel Level Gage
11. Fuel Tank
12. Intake-Air Duct
13. Command Radio Antenna
14. Fuselage Aft-Section Attaching Joint
15. Elevator Control Differential
16. Air Speed Pitot
17. Tail Pipe Support Track
18. Tail Pipe
19. Remote Compass Transmitter
20. Tail Pipe Clamp
21. Elevator Tab Motor
22. Engine
23. Intake-Air Seal
24. Engine Mounts
25. Aileron Booster Unit
26. Wing Spars
27. Aileron Torque Tube
28. Elevator Push-Pull Tube
29. Identification Radio
30. Sub-Cockpit Junction Box
31. Battery
32. Identification Radio Antenna
33. Elevator and Aileron Control Assembly
34. Nose Alighting Gear
35. Rudder Pedals
36. Fuselage Nose-Section Attaching Joint
37. Cartridge Case Ejection Doors
38. .50 Calbr Machine Guns (6)

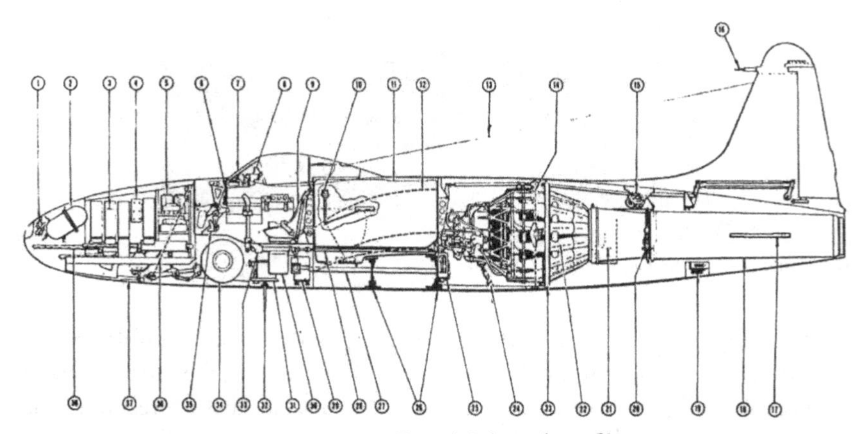

FIGURE 7.27. P-80A inboard profile.

commonly used for inlets on the side of a fuselage, but for the P-80A this fix was novel. We note that the prototype XP-80 "Lula Bell" did not have an inlet boundary-layer bleed duct.

By 1943, Kelly Johnson had bought into the concept of laminar flow airfoils, especially for use with jet-propelled airplanes. The P-80 had an

NACA 65-213 laminar flow airfoil. "With no propeller to develop a turbulent airflow over the wing," he wrote, "every opportunity exists to take full advantage of laminar flow airfoil design."[65] In regard to the design of the P-80, he stated "An extreme effort was made during the P-80A development to obtain aerodynamic smoothness over the whole airplane." The effort included rigid manufacturing tolerances and using the "best available putties, fillers, and paints to get smoothness." This was all to suppress the transition to turbulent boundary layers over the surface, and to maintain laminar flow over as long a distance over the whole airplane surface as possible. The aerodynamic smoothness did indeed give a speed increase up to 25 mph for cruising conditions and a noticeable benefit was obtained in range and climb. At high speeds, however, near the critical Mach number, the local shock waves that appeared on the surface triggered a turbulent boundary and in Johnson's words, "little is gained by extreme smoothness." This was a case of a "grand designer" following the best technology of the day but this time with mixed results.

The P-80 was the first United States operational jet fighter. A total of 1,715 fighter versions of the P-80 were produced. Spin-offs from the P-80 design were the T-33 jet trainer, the most widely used post-war jet trainer, and the F-94 Starfire, the Air Force's first all-weather jet fighter. On November 8, 1950, early in the Korean War, the P-80 made aviation history for participating in the first battle between jet fighters. With his design of the P-80, Kelly Johnson became even more of a "shooting star" within the Lockheed Aircraft Company and throughout the professional world of airplane designers in general. Johnson was well on his way to becoming an icon in his field.

DESIGNING THE SKUNK WORKS

All of the "grand designers" treated in this book designed airplanes. Kelly Johnson, however, was the only one of them to also design a unique group of engineers and technicians that functioned as a direct extension and magnification of his unique design talent – the Skunk Works. The Skunk Works was Kelly Johnson, and Kelly Johnson was the Skunk Works. Ben Rich, who much later succeeded Johnson as director of the Skunk Works after Johnson had retired, joined the group as a

[65] Ibid, p.669.

thermodynamicist in 1954. He aptly described the work environment and Johnson's persona in the Skunk Works at that time:[66]

All of us were well aware that we worked for Chief Engineer Clarence 'Kelly' Johnson, the living legend who had designed Lockheed's Electra and Lockheed's Constellation, the two most famous commercial airliners in the world. (Author's note: Here, Rich is being very parochial; the Douglas DC-3 designed by Art Raymond may best deserve such an accolade.)

Rich goes on to describe Johnson's presence in the Skunk Works:

All of us had seen him rushing around in his untucked shirt, a paunchy, middle-aged guy with a comical duck's waddle, slicked-down white hair, and a belligerent jaw. He had a thick, round nose and reminded me a lot of W.C. Fields, but without the humor. Definitely without that. Johnson was all business and had the reputation of an ogre who ate young, tender engineers for between-meal snacks. We peons viewed him with the knee-knocking dread and awe of the almighty best described in the Old Testament. The guy could just as soon fire you as have to chew on you for some goof-up. Right or not, that was the lowdown on Kelly Johnson. One day, in my second year on the job, I looked up from my desk and found myself staring right into the face of the chief engineer. I turned pale, then crimson. Kelly was holding a drawing of an inlet I had designed. He was neither angry nor unkind while handing it back to me. 'It will be way too draggy, Rich, the way you designed this. It's about twenty percent too big. Refigure it.' Then he was gone. I spent the rest of the day refiguring and discovered that the inlet was eighteen percent too big. Kelly had figured it out in his head – by intuition or maybe just experience? Either way, I was damned impressed.

That was the Skunk Works. Its organization and atmosphere was totally designed by Kelly Johnson. Moreover, there was no previous template for this, no previous "design" for Johnson to build upon. The Skunk Works was original in every respect. Moreover, Kelly Johnson did much to shield and protect the Skunk Works during the lean years between the end of World War II in 1945 and the beginning of the Korean War in 1950. Shortly after the outbreak of the Korean War, Kelly Johnson designed a new jet fighter that set aviation records and once again put the Skunk Works well ahead of the established state of the art.

THE F-104 STARFIGHTER

Johnson was always a great believer in getting out in the field and experiencing firsthand the environment encountered by new airplane designs,

[66] Rich and Janos, pp. 107–108.

beginning with his flight test experience with the Lockheed Electra in the mid-1930s. In 1952 he toured operational units of jet fighter squadrons in Korea, mainly to find out what increase in airplane performance was wanted by the pilots. The responses were predictable – greatly increased speed, faster rate of climb, and higher operational ceiling. When Johnson returned to the Skunk Works in December 1952, he had the conceptual design solidly in mind – a new, lightweight interceptor that could achieve sustained cruise at Mach 2.

At that time, aeronautical engineering had already entered the era of supersonic flight. On October 14, 1947, the rocket-powered Bell X-1 with Captain Chuck Yeager at the controls, became the first piloted airplane to break the speed of sound, achieving Mach 1.07. The X-1 was purely a research airplane; it was not designed to evolve into a future combat airplane. The same can be said about the Douglas D-588-2 Skyrocket, a swept-wing research airplane that was the first to achieve Mach 2, on November 20, 1953, with Scott Crossfield at the controls. Right on Crossfield's heels was Chuck Yeager flying the Bell X-1A, achieving Mach 2.44 on December 12, 1953. These airplanes and pilots broke the world of supersonic flight wide open. Keying on the obvious success of these airplanes, the NACA focused a major portion of its aerodynamics research program on supersonic aerodynamics.

With this technical environment in the background, after his return from Korea, Kelly Johnson was anxious to take a practical supersonic fighter design to the Air Force, a fighter that could achieve sustained cruise at Mach 2. That concept resulted in the Skunk Works next innovative design, the Lockheed F-104 Starfighter,

The Air Force already had a supersonic fighter ready for production, the North American F-100 (Figure 7.28). This airplane was a descendent of the earlier F-86 Sabre, and indeed was labeled the Super Sabre. We might be tempted to trace its roots all the way back to North American's grand designer, Edgar Schmued (Chapter 6). The F-100 was the world's first supersonic fighter, capable of Mach 1.39 at 30,000 feet. The airplane Kelly Johnson had in mind was considerably more advanced, an interceptor that would be capable of sustained Mach 2 flight.

From earlier pioneering NACA research on supersonic wing design by Walter G. Vincenti,[67] there evolved two extremes of supersonic airplane

[67] Vincenti, W. G., "Comparison between Theory and Experiment for Wings at Supersonic Speeds," NACA TR 1033, 1947.

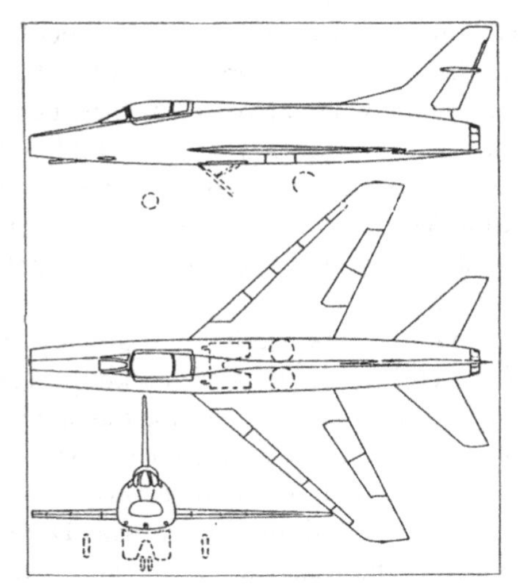

FIGURE 7.28. Three-view of the North American F-100D.

design. One used a highly swept wing of moderate thickness and aspect ratio, as exemplified by the F-100 shown in three-view in Figure 7.28. The other extreme is a straight, very low aspect ratio wing with an extremely thin airfoil section, as exemplified by the F-104, shown in three-view in Figure 7.29a, and in the photograph in Figure 7.29b. The major function of both extremes in design is to reduce the high aerodynamic drag caused by shock waves at supersonic speed – to reduce supersonic wave drag.

Johnson chose the second approach and designed a supersonic fighter that had a straight wing with the very low aspect ratio of 2.45. The airfoil was bi-convex in shape, a classic early supersonic airfoil shape that could be found in the most modern aerodynamic textbooks of the day. It was extremely thin, only 3.4 percent thick, with a very sharp edge – so sharp that when the airplane was on the ground, a glove was fitted over the leading edge to protect the ground personnel from being injured by the leading edge and to protect the leading edge from being injured by the ground personnel. In general the F-104 was the epitome of excellent classic supersonic airplane design, with a slender fuselage and pointed nose, as seen in Figure 7.29. The whole purpose of the F-104 was to fly at a sustained speed of Mach 2, and the design was aerodynamically optimized to achieve the lowest possible supersonic wave drag. Kelly Johnson

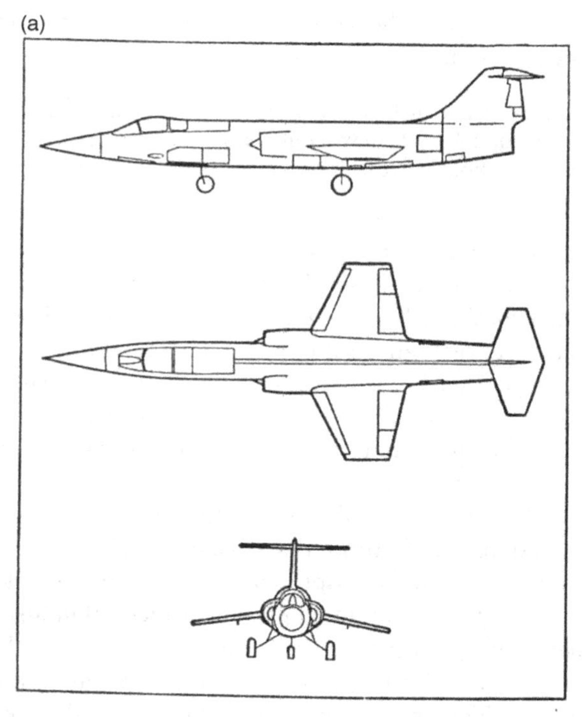

FIGURE 7.29. (a) Three-view of the Lockheed F-104.
(b) The Lockheed F-104.

had studied the supersonic aerodynamic literature available at the time, and clearly knew what he was doing.

Since Lockheed did not have a Mach 2 supersonic wind tunnel at the time, Johnson suggested a novel mode of testing the thin, short wings on the F-104. During a visit to the Air Force's Air Research and Development Command, he mentioned the following to the general in charge:[68]

> If we had a bunch of five-inch rockets, we could put the wing models on the rockets. If we shot enough of them we could find out how to make a wing with the thickness we want – about twice that of a razor blade. We can see if it flutters or not in supersonic flight.

Two weeks later, about 460 rockets showed up at the Skunk Works door, shipped by the Air Force from Korea. Different wing models were mounted on the rockets which were fired at Edwards Air Force Base.

The Air Force gave Johnson a contract for the F-104 on March 12, 1953. Almost a year later, the first test flight took place. The F-104 was unlike any airplane in the air, in its configuration and performance. On April 2, 1956, with test pilot Joe Ozier flying the F-104, the airplane exceeded Mach 2 for the first time.

Ultimately, the F-104 suffered from its extreme point design for Mach 2. All the aerodynamic features that made the airplane an excellent supersonic design were far from optimum when the airplane flew subsonically. It was prone to a well-known problem called "thin-airfoil stall" at low speeds, and the F-104 suffered a number of accidents on landing and takeoff. Also, the Air Force began to realize that most air combat did not take place at very high speed and altitude, but rather the combat arena was down around Mach 1 or less. The Air Force deployed F-104s for about a year and then consigned them to the Air National Guard. Kelly Johnson, always the entrepreneur, designed a heavier, all-weather, multirole attack version, the F-104G, specifically to satisfy the needs of the German Luftwaffe, which ultimately bought over 900 of these aircraft. The F-104G and later variants became widely used by other countries as well, such as Italy, Belgium, the Netherlands, Norway, Denmark, Turkey, Greece, Jordan, Pakistan, Formosa, Canada, and Japan. Later, NASA used almost a dozen F-104's for astronaut training.

The F-104 set new performance records. On May 1, 1958, it set the new world speed record of 1,404 mph and on December 14, 1959, it set the altitude record of 103,389 feet. In 1959, the prestigious Collier Trophy, awarded each year for the "greatest achievement in aviation," went to Kelly Johnson for the design of the F-104.

By then, Johnson had risen to vice president of research and development at Lockheed while still directing the Skunk Works as director of

[68] Johnson, Kelly: *More Than My Share of It All*, p. 109.

special projects. In spite of his growing administrative profile, Johnson remained an airplane designer at heart and in reality. He was on a roll.

THE U-2

In 1953, as the Cold War was ramping up, various leaders at Lockheed including Kelly Johnson, became aware of the vital national need for reconnaissance of the Russian's capability to launch intercontinental ballistic missiles (ICBMs) against targets in the United States. An airplane was needed that could fly well above the reach of the Soviet antiaircraft missiles and take photographs of Russian launch facilities on the ground. Johnson's first step was the well-established first step of conceptual airplane design – determine the requirements. In this case, the requirements were driven by simple necessity. The airplane had to fly above 70,000 feet, an altitude that could not be reached by existing Soviet ground to air missiles and above which no vapor trails could be visually observed to give away the airplane's presence. Moreover, as now becoming usual for new Skunk Works projects, there was no previous airplane of similar nature to serve as a starting point for a new design. Johnson and his small, close-knit staff once again designed a totally new jet aircraft that looked nothing like any seen before. The resulting design, labeled the U-2 ("U" for "utility") is shown in the three-view in Figure 7.30 and the photograph in Figure 7.30b.

The stunning aerodynamic feature of the U-2 is the very large aspect ratio of the wing that virtually leaps out of Figure 7.30, an aspect ratio of 14.3. By comparison, the aspect ratio for most conventional airplanes is on the average, around six to seven. The design aspect ratio of the wing of an airplane is a compromise between aerodynamics and structures, a classic compromise well understood by airplane designers since the 1920s. Aerodynamically, the induced drag coefficient (due to the presence of vortices generated principally at the wing tip and which trail downstream of the wing) is reduced by increasing the aspect ratio of the wing. Indeed, the induced-drag coefficient (sometimes called the vortex-drag coefficient) varies inversely with the aspect ratio; by doubling the aspect ratio, everything else being the same, the induced-drag coefficient is reduced by one-half. However, the required structural strength of the wing, and hence its weight, increases with aspect ratio. As a result of this compromise, airplane designers have conventionally settled on aspect ratios on the order of six to seven. (The Wright brothers fixed on an aspect ratio of 6.4 for the 1903 Wright Flyer – see Chapter 2.)

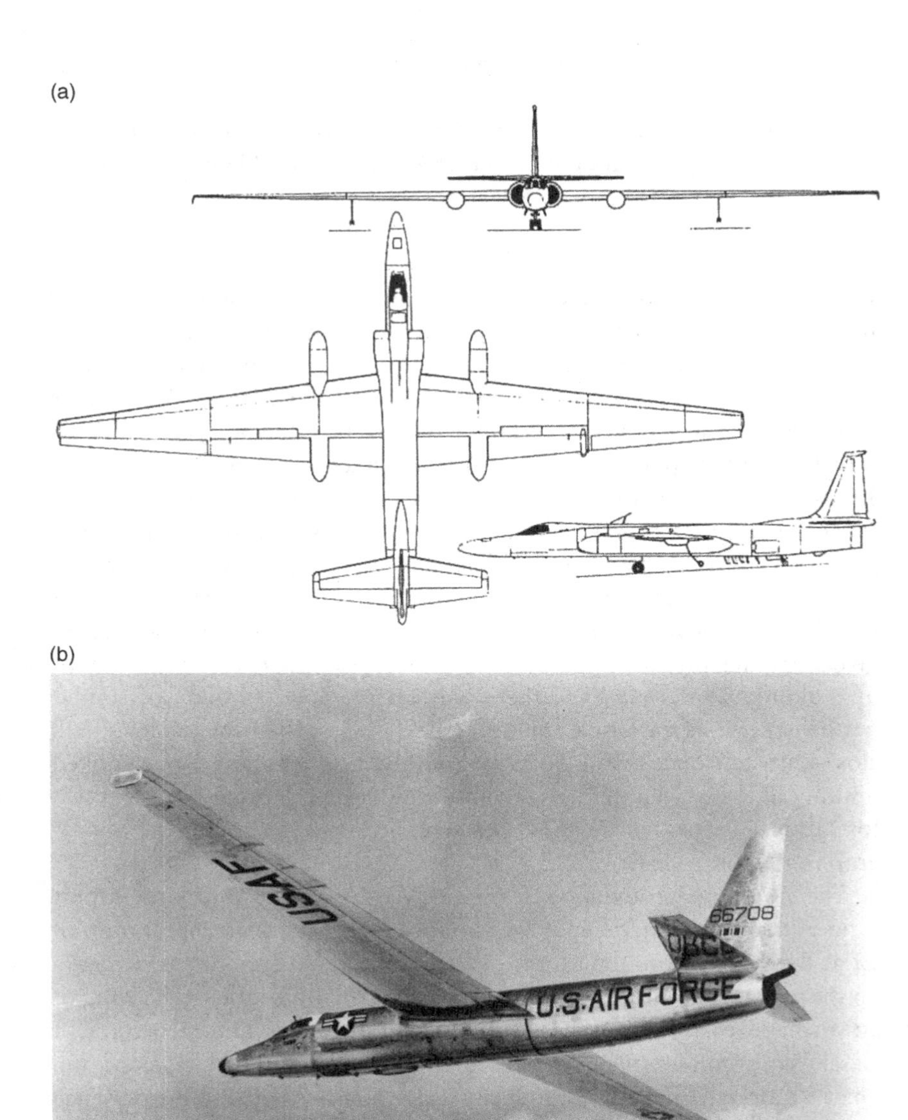

FIGURE 7.30. (a) Three-view of the Lockheed U-2.
(b) The Lockheed U-2.
(National Air and Space Museum, SI-80-8646.)

Kelly Johnson, ever since his education at the University of Michigan, fully understood the nature of this compromise. He stretched the bounds somewhat with this P-38 design, which had an aspect ratio of 8.26. The aspect ratio of his P-80 design was a more conventional value of 6.37. The flight performance requirements for the U-2, however, were well beyond those of any previous airplane and forced Johnson to design a wing which went well outside the bounds of the usual compromise. The problem was the very low atmospheric air density at altitudes of 70,000 feet and above. For an airplane to fly in steady, level flight, its lift must equal its weight. To generate the requisite amount of lift at 70,000 feet, a given airplane must fly fast (lift varies as the square of the airplane's velocity), and it must be at a high angle of attack (lift varies directly proportional to the angle of attack). At its required operational altitude, the U-2 flew fast, basically at its critical Mach number, the speed above which shock waves and the undesirable wave drag would occur. Also, at the same flight condition, the U-2 was at a very high angle of attack in order to generate a high lift coefficient. Indeed, at this flight condition, the U-2 was flying at a high velocity which was at the same time only a few miles per hour above its stalling speed due to its high angle of attack, a very challenging situation for the pilot. Superimposed on the high altitude requirement for the U-2 was another requirement, namely that for a long range to cover a great expanse of territory over the Soviet Union. Since the maximum range of a jet-propelled airplane varies inversely with drag coefficient,[69] Johnson knew that the U-2 design must be driven by the aerodynamic requirement to minimize the drag coefficient. He also knew that at very high angles of attack, the drag of an airplane is predominantly induced drag, and that the induced drag coefficient varies as the *square* of the lift coefficient. Double the lift coefficient and the induced drag coefficient increases by a factor of four. Faced with this situation, Kelly Johnson had only one real solution. Since the induced drag coefficient varies inversely with aspect ratio, the only way to design the airplane to have an acceptably small enough drag was to have the wing aspect ratio absolutely as high as possible in spite of the increased structural weight penalty. This is why the airplane shown in Figure 7.31 has the extraordinarily large aspect ratio of 14.3. Kelly Johnson, being a "grand designer" in the spirit of this book, designed the U-2 using the methodology of conceptual airplane design pioneered by Frank Barnwell (Chapter 3), but

[69] Anderson, *Introduction to Flight*, 7th ed., p. 509.

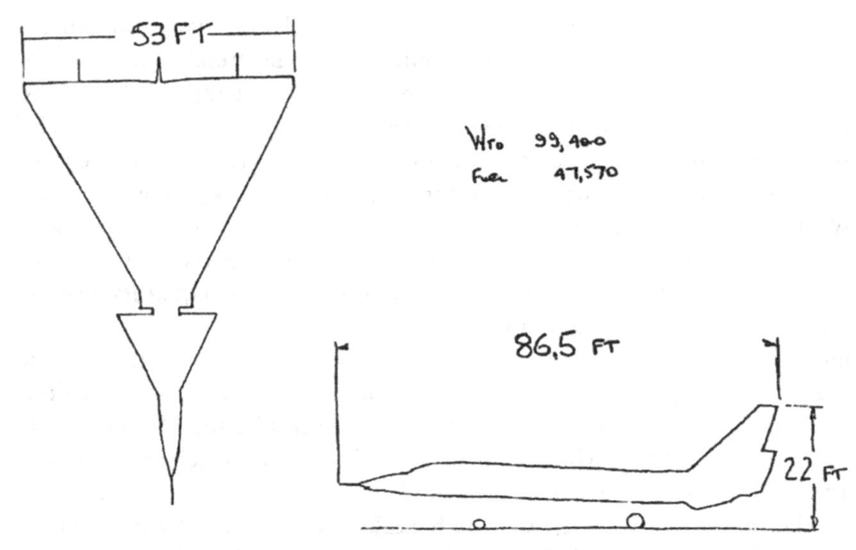

FIGURE 7.31. Johnson's conceptual sketch of the SR-71; side view, 1958.

was willing to go beyond the accepted technical limits for wing design in order to ensure the success of his airplane.

And indeed the U-2 was a success. Its official first flight took place with great secrecy on August 8, 1955. Everything about the U-2 was wrapped in the highest secrecy. The Skunk Works produced this airplane not for the Air Force, but rather for the Central Intelligence Agency which controlled the missions. U-2 pilots carried out their missions over Russia with impunity for four years before the Russians developed a missile and radar system capable of shooting down the airplane. On May 1, 1960, Francis Gary Powers, flying a U-2, was shot down by a Russian missile over the Sverdlovsk complex. Powers survived and became a high-profile prisoner in Russia before finally being exchanged in February 1962. President Eisenhower was forced to admit the U-2 spy operation, and the overflights of Russia ceased. However, the U-2, because of its unique mission requirements and performance, continues to be used even today by the US Air Force for reconnaissance as needed, and by NASA for high-altitude research.

SR-71 BLACKBIRD

"We got nailed over Sverdlovsk by an SA-2. That's that, we're dead," Kelly Johnson told an assembled group at the Skunk Works the day after Gary

Powers was shot down in his U-2. It was the first time that an airplane had been shot down by a ground to air missile. To Johnson, however, this was no surprise. As early as 1958, with two years of spectacular success already with the U-2, Kelly Johnson felt the airplane was living on borrowed time. He shared with a small group of colleagues that a successor to the U-2 was desperately needed, a new airplane that could cruise at Mach 3 at 90,000 feet – much faster and higher than the U-2.

Johnson was dead serious. On April 21, 1958, he took out four sheets of blank paper and covered them with longhand conceptual design calculations for such an airplane. These calculations, of I have a copy, are a classic case of conceptual design in the spirit of Frank Barnwell. The first words that Johnson wrote down were a list of *requirements*, step one of the conceptual design process:

High Altitude Cruise – 90,000 feet
Design Cruise Mach No. -3.0
Engines – two
Crew – Basic – one (Two in future)
$\frac{L}{D}$ reqd. – 7 to 8
Range – 2,000 mile radius
Payload – 500 lbs.

On the second page, he chose two Pratt and Whitney J-58 jet engines, each capable of producing a thrust of 4,000 pounds flying at 90,000 feet and at Mach 3. Since in level flight, thrust equals drag, the airplane's drag at the stated flight requirements is 2 x 4,000 = 8,000 lbs. One of the requirements Johnson stipulated was a lift-to-drag ratio between 7 and 8. On the second page of his calculations, he picks L/D = 7.5, which immediately fixes the total airplane weight equal to 7.5 x 8000 = 60,000 lbs. This is the combined weight of the airframe, engines, and fuel. Using the well-known equation for range of an airplane,[70] Johnson calculates the weight of fuel required for a range of radius 2,000 miles, to be 43,400 lbs. He follows this figure with *two exclamation marks*, because he already knows this is a substantial portion of the total allowed weight of 60,000 lbs. He remarks on the third page that the fuel plus engine weight is 43,400 + 12,000 = 55,400 lbs. This would allow only 4,600 lbs. for the whole airframe plus payload weight, or "80% of U-2" as he writes neatly on page three of his calculations. In the true spirit of

[70] Ibid, pp. 506–510.

the conceptual design methodology, over the next two months he plays with different options for the airframe weight and has two members of his staff come up with a more detailed weight breakdown accounting for items such as landing gear, controls, hydraulics, electronics, etc. At the end of a five-page calculation dated June 26, 1958, that now deals with a much heavier airplane of 100,000 lbs. gross, Kelly Johnson writes and underscores at the bottom of the page "*our only chance*."

These hand calculations, made by Johnson on the same, now worn slide rule he had at the University of Michigan, are a striking example of the same intellectual methodology for conceptual airplane design as used by Frank Barnwell in 1916 for strut-and-wire biplanes. Except in 1958 Kelly Johnson is dealing with a Mach 3 jet airplane to fly at 90,000 feet, pressing the bounds of existing advanced technology. The airplane that eventually was to emerge from these early calculations was the SR-71 "Blackbird"; although many of the characteristics of the SR-71 are still secret, the eventual weight of the actual airplane is estimated to be about 145,000 pounds.

Making a sketch, frequently by hand, of the general configuration of the airplane is one of the steps in conceptual design (see Figure 2.7). On June 26, 1958, Kelly Johnson drew on a sheet of standard size paper a top and side view of the configuration he had in mind (Figure 7.31). On a second sheet of paper, he drew a second top and side view, this time sketching the internal placement of the jet engines, wing fuel tanks, a radar dish in the nose, electronics packages (Johnson labeled it on the drawing simply as "black boxes"), and a hold for missiles inside the middle of the fuselage (Figure 7.32). This was the first conceptual sketch of the airplane that eventually evolved into the SR-71 and it was drawn by Johnson as one of the normal steps of the methodology of conceptual airplane design originally set forth by Frank Barnwell in 1916.

Johnson logically identified his conceptual design as the "U-3," so labeled on the sheets of his calculations. The airplane was intended to be the successor to the U-2 as a much faster and higher flying reconnaissance vehicle. The label U-3, however, was not an official designation. The design later went through a series of modifications; the twelfth such design was given the company label the A-12. Noting in Figure 7.32 that Johnson made a provision for a missile bay in the fuselage, clearly some intent was for the airplane to also serve as a very high performance interceptor. Later designated the YF-12, this version did not come to fruition, primarily because of a diminished threat from Soviet bombers. It was, however, the version from which the SR-71 was derived. The

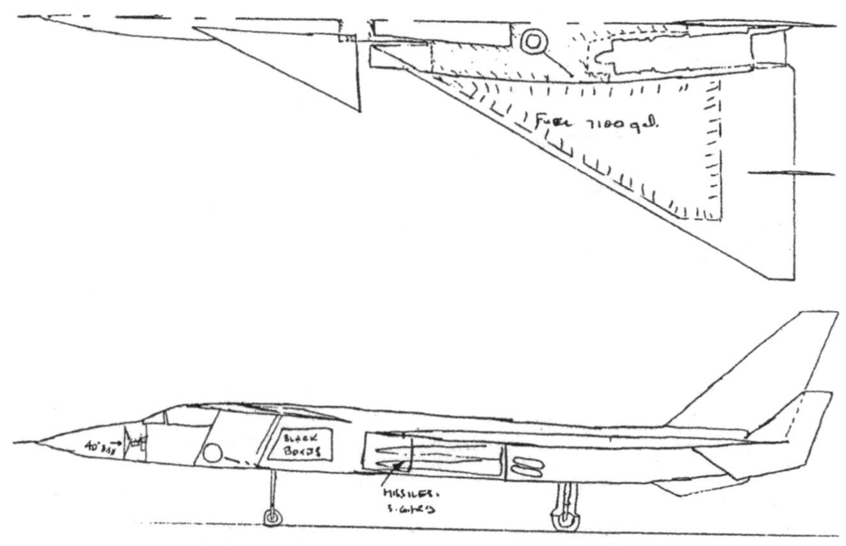

FIGURE 7.32. Johnson's conceptual sketch of the SR-71; top view, 1958.

SR-71 is shown in Figure 7.33. Although many modifications later, the SR-71 still reflected the basic essence of the "U-3" in Kelly Johnson's conceptual design sketches. The design, however, had grown considerably in size. The SR-71 had a length of 107 feet, 5 inches, a wing span of 55 feet, 7 inches, and the weight had grown to 152,000 pounds.

The SR-71 is a crowning example of how the intellectual methodology of conceptual airplane design basically had remained the same since Frank Barnwell had first quantified it in 1916, and that the exponential advancement of the airplane in the twentieth century was due to the courage and foresight of "grand designers" being willing to hang new technology on their designs. This was extreme in the case of Johnson's design of the SR-71. He later wrote:[71]

> The idea of attaining and staying at Mach 3.2 over long flights was the toughest job the Skunk Works ever had and the most difficult of my career. Early in the development stage, I promised $50 to anyone who could find anything easy to do. I might as well have offered $1,000 because I still have the money.

He went on to emphasize the extreme technical challenges:

[71] Johnson, Kelly: *More Than My Share of It All*, p. 137.

(a)

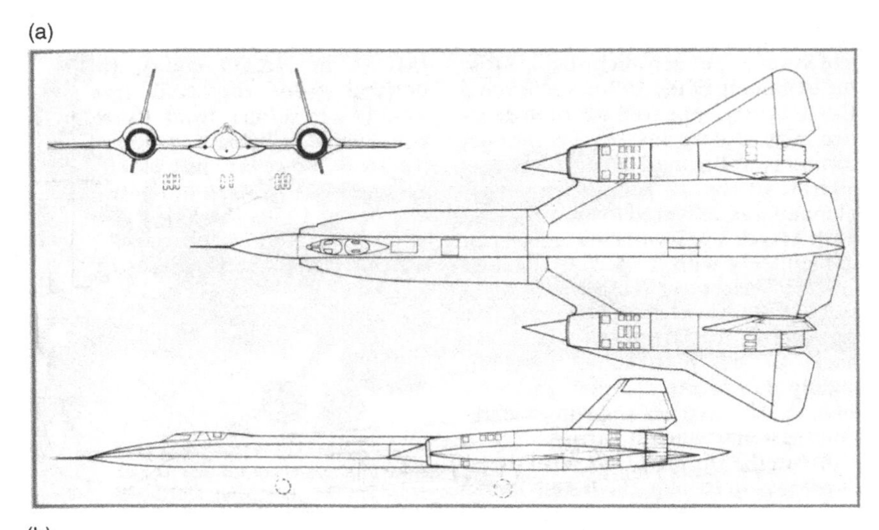

(b)

FIGURE 7.33. (a) Three-view of the SR-71 Blackbird.
(b) The Lockheed SR-71 Blackbird.
(NASA.)

Aircraft operating at those speeds and altitudes would require development of special fuels, structural materials, manufacturing tools and techniques, hydraulic fluid, fuel-tank sealants, paints, plastics, wiring and connecting plugs, as well as basic aircraft and engine design. Everything about the aircraft had to be invented. Everything.

At a press conference on July 24, 1964, President Lyndon Johnson announced publically the existence of a "new spy plane," the SR-71, to be employed by the Strategic Air Command. Later that year, on December 23, the Department of Defense announced that the first flight of the US Air Force SR-71 had taken place the day before. Kelly Johnson's new creation had become public knowledge.

The first record of the SR-71 is still cloaked in secrecy to a certain extent. The record shows, however, that thirty-two were built by Lockheed at its Burbank factory. The airplane served in the US Air Force beginning in 1964 until its final retirement in 1998. During this period it smashed all conventional flight records. On July 28, 1976 an SR-71 set the altitude record for steady level flight of 85,069 feet. On the same day another SR-71 set the speed record of 2,193.2 mph. In terms of long distance flights, on September 1, 1974, an SR-71 set a record of one hour and fifty-five minutes from New York to London. A transcontinental record was set when an SR-71 took off from Burbank (its home) and landed at Dulles Airport in Washington, DC sixty-four minutes and twenty seconds later. That SR-71 taxied to a storage space to await eventual display at the Smithsonian's National Air and Space Museum Udvar-Hazy Center. All these records still stand. The SR-71 still remains the fastest and highest flying airplane with air-breathing engines in history and at the time of writing, there is no other aircraft on the horizon that can do better. The SR-71 was the jewel in Kelly Johnson's design crown. Today, exactly one-half of the SR-71s that were produced are in air museums in the United States and Europe. Most of the others are proudly displayed by the US Air Force at air bases around the country.

When Kelly Johnson finished his conceptual design for the "U-3" on four pieces of paper using his slide rule (which he jokingly referred to as his "Michigan computer" since he had been using it since his days at the university), the long line of "grand designers" as typified by the case histories used in this book also came to an end. The intellectual methodology of conceptual airplane design had not changed but the process of hanging new technology on the design certainly had. Rather than the almost unilateral process carried out by the "grand designer" that we have seen previously, Kelly Johnson now had to rely on a growing number of experts in the Skunk Works. For example, Ben Rich, a mechanical engineer who Johnson had brought into the Skunk Works as an expert thermodynamicist, was given the enormous responsibility of developing the engines – an absolutely new technical challenge brought on by the Mach 3 plus sustained speed of the SR-71. Ultimately, the new hybrid gas turbine and ramjet engine developed by Rich and his team became a part of the success story of the SR-71. The engine had a Pratt and Whiney J-58 gas turbine as an internal core, wrapped inside a larger cowling with a flow passage around the J-58. For takeoff and lower speed flight, the J-58 provided the thrust. At the higher Mach numbers, however, the air flow from the cowling inlet was shunted around the J-58 and burned with fuel in an afterburner – essentially operating as a ramjet.

Rich was also responsible for the airplane being painted black. At Mach 3, aerodynamic heating of the external surfaces of the airplane was severe, creating surface temperatures too high for the use of aluminum conventional for other, lower speed aircraft. Instead, Johnson and his design team had to use titanium for the external skin of the airplane, a difficult material to machine. Ben Rich tells of the following interaction with Johnson when Rich suggested a means to allow the use of a softer titanium, good for machining but weaker at higher temperature.[72]

> My idea was to paint the airplane black. From my college days I remembered that a good heat absorber was also a good heat emitter and would actually radiate away more heat than it would absorb through friction (aerodynamic heating). I calculated that black paint would lower the wing temperatures 35 degrees by radiation.

But Johnson hit the ceiling. Airplane designers always have an aversion to adding weight to a given design. He exclaimed to Rich that "you're asking me to add weight – at least a hundred pounds of black paint – when I'm desperately struggling to lose even an extra ounce. The weight of your black paint will cost me about eighty pounds of fuel." When Rich tried to explain that the airplane would be much easier to build using the softer titanium, which the lower skin temperature would allow, Johnson retorted: "Well, I'm not betting this airplane on any damned textbook theories you've dredged up. Unless I got a bad wax buildup, I'm only hearing you suggest a way to add weight." By the next day, however, Johnson had changed his mind. "On the black paint," he told Rich, "you are right about the advantages and I was wrong." This was a rare

[72] Rich and Janos, p. 203.

admission from Johnson. He approved the idea and beginning with the first prototype that rolled out of the Skunk Works, the SR-71 became known as the Blackbird.

To the very end of his design career, Johnson continued his penchant for publishing in the open technical literature. In 1969, he gave a paper to the American Institute of Aeronautics and Astronautics entitled "Some Development Aspects of the YF-12A Interceptor Aircraft."[73] This title was for all practical purposes, a euphemism for the SR-71, but even Kelly Johnson had to tiptoe around very tight military security restrictions. He covered in a general way the new problems encountered with using titanium rather than aluminum. He listed electronic systems, radomes, fuels, and lubricants as subsystems requiring special attention in the high aerodynamic heating environment. He once again noted perhaps the overarching technical challenge for the SR-71: "I believe I can truly say that everything on the aircraft from rivets and fluids, up through materials and power plants had to be invented from scratch." He noted that the number of engineers in the design effort was "considerably less than 200 at its peak," but that "these were very experienced personnel who performed well in producing the advanced weapon system." Ben Rich later noted that the entire Skunk Works design team for the SR-71 was seventy-five. Although this number is small compared to the massive design teams used today by large companies for new airplane designs, it is, by comparison with our previous test cases, a larger number of people assisting a classic "grand designer."

Another change was taking place in Kelly Johnson's technical modus operandi, dictated by modern circumstances. As much as he enjoyed writing to inform his reading public about his work, unlike his long and detailed paper in 1947 in the *Journal of the Aeronautical Sciences* on the design of the P-80 in which he had real engineering data and numbers, his 1969 AIAA paper contained *very little hard data and few numbers*. The world of airplane design had become, even more than usual, shrouded in secrecy. His paper did, however, reveal an interesting fact about the engine. In a graph showing the *percentage* of the thrust being contributed by various parts of the engine – conical entrance cone, cowling, inlet, the J-58 gas generator, and the afterburner exit – over 70 percent of the thrust

[73] Johnson, Clarence L., "Kelly" Johnson, "Some Development Aspects of the FY-12A Interceptor Aircraft," AIAA Paper 69–757, 1969; reprinted in *Perspectives in Aerospace Design*, compiled by Conrad F. Newberry, American Institute of Aeronautics and Astronautics, Reston, VA 1991, pp. 129–136.

at high speed is generated by the internal air pressure distribution exerted on the cowling *inlet* alone, and not from the J-58. Johnson crowed: "My good friends at Pratt and Whitney did not like me to say, that at high speeds, their engine is only a flow inducer, and that after all, it is the nacelle pushing the airplane."[74]

In 1964, Kelly Johnson received his second Collier Trophy, this time for the design of the YF-12A interceptor; the SR-71 was still under strict security wraps. The award stated that Johnson's achievement was the greatest in American aviation.

RETIREMENT

Kelly Johnson retired from Lockheed in 1975 and Ben Rich, handpicked by Johnson, became the second director of the Skunk Works. A man such as Johnson, however, does not intellectually retire. He called it "partial retirement." He continued to consult on various projects, becoming less involved with time. His personal life, however, became a sequence of ups and downs. His wife of thirty-two years, Althea, died of cancer in December 1969. He piloted a small airplane over Santa Barbara Bay and following her wishes, scattered her ashes in the sea. In May 1971, he married Maryellen Meade, his secretary, and they experienced a very happy life together until Maryellen died of a protracted illness in October 1980. Before her death she had encouraged Kelly "not to remain alone." One month later Johnson married Nancy Harrigan, who had been a good friend of the Johnsons.

In 1986, Kelly Johnson fell and broke his hip. He was admitted to St. Joseph's Hospital, very close to the Lockheed offices in Burbank. He stayed there until his death four years later. Although his broken hip gradually mended, he experienced a long period of physical deterioration and rapidly progressing dementia. This man, who had spent his life as a consummate aeronautical engineer and airplane designer, arguably the grandest of the "grand designers," spent the last four years of his eighty-year life in virtual hell. Kelly Johnson died on December 22, 1990, ironically exactly the same day that Ben Rich retired as head of the Skunk Works. The following day, Lockheed ran a full-page black bordered tribute in the *Los Angeles Times*, simply showing the Skunk Works Logo with a single tear rolling down its cheek.

[74] Ibid, p. 135.

REFLECTIONS ON KELLY JOHNSON AS A GRAND DESIGNER

Kelly Johnson strongly believed in himself, and for good reason. He was rarely wrong. At the Skunk Works he offered a quarter to any of his staff who could prove him wrong in a technical matter; he gave out very few quarters. He was exceptionally driven from the time he was 12 years old and wanted to be an airplane designer to the very end of his life. He was good, and he knew it. He studied hard and learned fast. His excellent education in aeronautical engineering at the University of Michigan put him in high gear, racing to one of the most illustrious careers in the profession. Aeronautical engineering and airplane design *were* his life.

In my opinion, however, there were at least two areas in which Johnson's almost dogmatic attitude put him on the wrong side of history. One was his total disregard for the value of the manned X-airplane series, a sequence of manned aircraft designed and flown purely for research purposes beginning in the 1940s with the Bell X-1, the first piloted airplane to break the speed of sound. Indeed, Chuck Yeager's flight in the X-1 on October 14, 1947, which achieved Mach 1.06, opened the world of supersonic flight; it was arguably the most significant event in the history of flight since the Wright Flyer successfully flew on December 17, 1903. The quest for speed and altitude carried the manned X-airplane series into the hypersonic regime with the successful X-15 hypersonic airplane program which spanned the 1960s, and generated a large data base for hypersonic flight which was essential for the later design of the space shuttle. Kelly Johnson was a member of the NACA Committee on Aerodynamics which met on October 5, 1954, to make a final decision on the launching of a manned hypersonic airplane program for research. Johnson was the only dissenting member of the committee, arguing that the previous manned research airplanes had provided "generally unsatisfactory" experience that could not benefit airplane companies designing practical supersonic airplanes. Despite Johnson's objections, the committee officially recommended the construction of a hypersonic research airplane, opening the door to the X-15. But that door did not help Kelly Johnson. Fifteen years later, Johnson made his feelings known to the professional aerospace engineering world in some veiled comments in his AIAA paper on the development of the YF-12A. Commenting on previous high-speed airplane data and its lack of applicability to the YF-12A, Johnson wrote:[75]

[75] Ibid, p. 129.

The North American B-70 was in its design stages. It was expected that a large amount of fall-out would result from this program and the NASA tests on the X-15. Both of these conditions did not apply however; the YF-12A rapidly passed the development status and took different paths than followed for the B-70. The X-15 with its very short duration of flight, even though at higher Mach numbers and altitudes, did not encounter the problems of air breathing power plant inlet design, ejector or steady state temperature conditions.

Here, Johnson is making the point that all the previous high-speed manned X-airplanes had been powered by rocket engines, whereas the YF-12A and the follow-on SR-71 were powered by air-breathing jet engines. He is making a second point that because of the very limited duration of the liquid-fueled rocket engines, the flights of the X-15 at hypersonic speeds were measured in minutes. He went on to elaborate:

In fact, in terms of cooling of the cockpit, the problem turned out to be at least seven times as hard on the YF-12A due to the steady state heat flux, than it was for the X-15. It is also true that, in the whole series of research aircraft from the X-1 through the latest types beyond the X-15, there were no power plant problems even remotely resembling those we encountered on the YF-12A. Most of the high speed X-series aircraft were either rocket powered, or followed conventional design, current at the same time on Military fighter aircraft.

As a "grand designer," Kelly Johnson as far back as his days at the University of Michigan had always favored research, but as brought out in his feelings about the series of research X-airplanes, primarily research that provided data that could help him design new airplanes.

Johnson's negative view of the NACA is the other aspect that put him on the wrong side of history. From its beginning in 1916 to its amalgamation into NASA in 1958, the NACA had been held in an almost iconic status by aeronautical engineers worldwide. In particular, the NACA was essential to airplane designers with its work on the NACA series of airfoils, including the pioneering laminar flow airfoils, and with the stunning, drag-reducing cowling design for air-cooled piston engines. Johnson readily used these and other NACA-produced data for his airplane designs during the 1930s. Things came to a head, however, with the P-38 tuck-under compressibility problem discussed earlier. Johnson clearly faulted the NACA for "contributing nothing to solving the problem of compressibility on the P-38 except allowing the use of its wind tunnel."[76] Throughout the 1930s and into World War II, however, the

[76] Johnson, Kelly: *More Than My Share of It All*, p. 71.

NACA had become the world's research leader on the understanding of the aerodynamic causes and consequences of the compressibility problem.[77] It was the NACA, through its research engineers with this knowledge that came up with the solution to the P-38 tuck-under problem, namely the small flap under the wing. But Kelly Johnson did not see it that way. In his mind the research carried out in the NACA high-speed wind tunnel was Lockheed's work, principally his own. In his autobiography, Johnson reflected:[78]

> Because of the importance of compressibility as an aviation industry problem and the wide interest in it, I prepared a technical paper covering our own research and what we thought the solutions might be for presentation to the American Institute of Aeronautical Sciences. It was duly cleared by the War Department, and I presented it at a meeting in January 1943. Naturally, there were many requests from other companies for copies and I supplied them.
>
> Then the paper was recalled and labeled secret.

The NACA version of this matter is clearly stated by Smith J. DeFrance, engineer in charge (director) of the NACA Ames Aeronautical Laboratory at Moffett Field, California, in a strong letter he wrote to Dr. George Lewis, then the director of aeronautical research for the NACA. In this letter he railed against Johnson:[79]

> Through the grapevine, a representative of one of the aircraft companies on the Coast has been informed that the Lockheed company installed an auxiliary flap approximately at 33 percent of the chord of the center section of the P-38 wing and flight-tested the combination in high-speed dives. The information I have is that Colonel Kelsey made the flight test and that after reaching an uncontrolled condition in a high-speed dive he possibly became frantic and operated the flaps suddenly with full deflection. The flaps tore off from the wing, probably struck the tail surfaces, and the tail disintegrated. Fortunately, Colonel Kelsey was able to bail out, but the airplane, which was the Lockheed company's special test plane, was lost with all of their flight instruments.
>
> Kelly Johnson still thinks apparently that the auxiliary flaps are the answer to his uncontrolled dive condition on the P-38 and it is understood that he is making another installation for further flight tests.

[77] Anderson, John D., Jr., "Research in Supersonic Flight and the Breaking of the Sound Barrier," in *From Engineering Science to Big Science*, ed. by Pamela E. Mack, NASA SP-4219, 1998, Chapter 3, pp. 59–90.

[78] Johnson, Kelly: *More Than My Share of It All*, p. 78.

[79] Letter from Smith J. DeFrance to George Lewis, April 26, 1943, NASA History Office record number 2967, Folder Lewis, George W.

To this point DeFrance is simply relating an experience that is totally consistent with Kelly Johnson performing as a "grand designer," clearly being willing to apply new technology to his design. But then DeFrance's letter turns combative. He continues to write:

> I think everything possible should be done to stop the Lockheed company from obtaining a patent on the auxiliary flap arrangement because in the first place, it is not Lockheed's idea, and in the second place, the primary development was carried out by the Committee.

The "Committee" he refers to is the National Advisory Committee for Aeronautics, the NACA. DeFrance then goes on to state: "As I told you some time ago, I got the idea of the auxiliary flap while reading the German report which commented on the adverse effect of the flap on the lower surface of the wing as a dive break." Then Smith DeFrance's real feelings came out as he ends his letter to Lewis as follows:

> Kelly Johnson apparently is not satisfied with having lifted the data from the 16-foot tunnel tests on the P-38 and used them in his paper before the Institute of Aeronautical Sciences, but now he would like to patent the work of the Committee. Unless it is stopped, the next thing, he will be trying to make the Committee a subsidiary of the Lockheed company.

Clearly, there was no love lost between the NACA and Johnson – on either side. This author knows of no other important airplane designer during the first half of the twentieth century with such negative feelings toward the NACA. Indeed, part of the original 1915 charter of the NACA was "to supervise and direct the scientific study of the problems of flight, with a view to their practical solution." The NACA carried out a great deal of research and development in support of the aircraft industry; indeed, it held an annual conference during the 1930s at the Langley Aeronautical Laboratory for members of industry and government during which it shared its research results and solicited suggestions for work that the NACA could and should be doing for the "practical solutions" of problems of flight. During World War II especially, the NACA focused on work aimed almost exclusively at improving the performance of existing US aircraft. This author knows of no situation where the NACA willfully hindered the work of airplane designers; indeed, the NACA was there to help.

A case might be made that Kelly Johnson, however, did not need the NACA. Of course he used NACA airfoil shapes on the P-38 and P-80 – virtually every airplane designer did in those days. But when it came to design data necessary for his creation of the F-104 and the U-2, Johnson obtained it himself with the help of his small, highly qualified staff in the Skunk Works. This work ethic came to a crescendo with the design of the SR-71, where everything was new, where Johnson and his staff had to "invent everything." Johnson was in effect, his own NACA – self-sufficient and self-confident. A "grand designer" of the magnitude of Kelly Johnson could do this.

Johnson's opinions about the high-speed research X-airplanes are somewhat understandable given his absolute concentration on the challenge of new airplane designs for practical, operational missions. This is indeed a mark of a "grand designer." On the other hand, his frustration with the NACA that came to a head with the P-38 tuck-under compressibility problem, in my opinion, was more a product of his dogmatic and get-it-done-no-matter-what personality rather than a fundamental fault of the NACA. The personalities of each of the grand designers in this book were all different. There is no common personality factor that is shared by all. Kelly Johnson's personality was simply part of his whole being, and it was his whole being that drove him to be a grand designer.

Indeed, Kelly Johnson was arguably the grandest of the "grand designers." He exemplified the three important aspects necessary to be highly successful in any endeavor: intense desire to be what you want to be; the wherewithal to attain your goals; and finally the native ability and intelligence to do it. From the age of 12 he developed an intense desire to be an airplane designer, and for the rest of his life his focus never wavered from this goal; for all practical purposes, he devoted virtually every minute of his being to aeronautical engineering, either in action or in thought. Fortunately, he had the wherewithal to attend the University of Michigan which had the most viable undergraduate program in aeronautical engineering at that time and a graduate program known for its excellence. Johnson took advantage of both programs. Also, his hiring by Lockheed during the Depression when many airplane companies were going under gave him the wherewithal to immediately take part in the development of the Lockheed Electra; essentially, Johnson got in on the ground floor of a newly reformed company, and never left it. Finally, Johnson excelled intellectually; he was smart and he loved learning; he was by nature equipped to take full advantage of the opportunities given him. And this he did, as attested by the long line of innovative and history-making Lockheed aircraft that flowed from his work. Thus, we end our limited parade of "grand designers" in this book with Kelly Johnson, arguably the *grandest* of the "grand designers."

8

Airplane Design and the Grand Designers

The End Game

This book posits that the stunning and exponential improvement in the airplane during the twentieth century was *not* due to any corresponding exponential growth and change of the intellectual methodology of *conceptual* airplane design, which has in its essence stayed about the same as outlined by Frank Barnwell in 1916, but rather it was due to the knowledge, willingness, and sometimes bravery of a given airplane designer to hang new and only to a certain extent proven technology. All the "grand designers" did this. In the early part of the twentieth century, the new technology was frequently incorporated directly into the conceptual design process itself, as was done by the Wright brothers with the 1903 Flyer, Barnwell with the Bristol fighter, and Art Raymond with the DC-1, 2, and 3 series. In later times the new technology evolved simultaneously with or after the conceptual design, and was incorporated later in the detailed design process. Witness the change from a standard NACA airfoil to a laminar flow airfoil by Edgar Schmued for the P-51, based on wind-tunnel tests after the conceptual design had been finished. The extreme example was Kelly Johnson during the evolution of the SR-71 Blackbird, for which the Skunk Works was constantly developing new technology out of necessity, after the conceptual design was finished.

In any event, the intellectual methodology of conceptual design as originally set forth by Frank Barnwell stayed essentially the same. What we see as the exponential improvement of the airplane is really the exponential improvement of the *technology* of the airplane. Over the first half of the century, piston engines rapidly became more reliable and powerful and jet engines came on the scene, providing a considerable increase in the *power available*, a critical factor in the conceptual design of airplanes for transonic and supersonic flight. Advances in aerodynamics progressively led to continual reductions in drag, hence to corresponding reductions in the *power required*, the other critical factor in the conceptual design process. The difference between power available and power required is defined as the *excess power*; everything else being equal, the larger the excess power, the faster the airplane can fly, the higher it can fly, or both. The rapid increase in excess power over the twentieth century was a technology development that led to exponential increases in airplane speed and altitude – factors predicted using the tried and proven methodology of conceptual design laid out by Frank Barnwell in 1916. All the "grand designers" understood and practiced this methodology.

We might ask the question *why* the intellectual methodology of conceptual airplane design did not experience an exponential improvement and change in parallel to the technology. The answer is rooted in the fact that the flight performance of an airplane follows the basic laws of nature. Technology changes and improves, but the laws of nature stay the same. The intellectual methodology of conceptual airplane design is simply an application of Newton's classic laws of motion. Of course, specific airplane designers use their own ingenuity and experience in applying these laws. For a given requirement, ten different designers will finally come up with ten different airplanes to satisfy these requirements, all of them obeying the same laws of physics. One of the aspects that distinguishes "grand designers" from others, however, is a particularly high degree of ingenuity, and an innate ability to simply "know" what will work.

How do "grand designers" come to this degree of knowing? All of them discussed in this book shared some common traits:

(1) All of them were totally absorbed in their work, to the virtual exclusion of much else. The Wright brothers were always dedicated to work, beginning with their bicycle business, and continuing with increased dedication to their flying machine activities. Neither Wilbur nor Orville married. Frank Barnwell designed over a hundred airplanes. Art Raymond devoted his time to designing a series of pioneering (and money-making) airplanes for Douglas. R. J. Mitchell, designing virtually *all* the Supermarine aircraft during the 1920s and 30s, including the famous Schneider cup racers and the still iconic Spitfire, the latter while he was suffering from terminal cancer. Edgar Schmued became a workhorse for North American Aviation, responsible for pacesetting fighter designs such as the P-51 Mustang (arguably the best fighter from World War II),

and the F-86 Sabrejet of Korean War fame, America's first swept wing jet fighter. Kelly Johnson started work as a child to help support his family, and continued practically nonstop through his education at Michigan, and through his career as Lockheed's (and arguably America's) most famous airplane designer.

(2) All of them were "turned on" by airplanes. Once the Wrights became interested in heavier-than-air flying machines in 1896, they hardly did anything else for the rest of their lives. Frank Barnwell was caught by watching airplanes flying as a young boy, and designing airplanes became his life's work from World War I to his untimely death in 1938 in a homebuilt airplane of his own design. Art Raymond knew what he wanted to do since his early aeronautical engineering education at MIT, and continued doing it for the rest of his long life. R. J. Mitchell similarly was motivated by the fragile bi-planes he saw flying over his hometown as a boy, and made up his mind to join an airplane company as soon as the opportunity presented itself at the end of World War I. Edgar Schmued became so interested in airplanes that he eventually left Germany between the two world wars, with its severely limited aeronautical activities dictated by the Allies, and moved to the United States to find much greater opportunity to be a designer of airplanes. Kelly Johnson kept a detailed scrapbook on airplanes as a young boy, and at the age of 12 knew he wanted to be an airplane designer. All the "grand designers" shared a passionate love of airplanes acquired when they were young.

(3) Due in part to their intense interest in airplanes, each grand designer eagerly read the latest aeronautical publications. For success, they all had the "need to know" about the latest developments in aeronautical technology. They did more than just read. They were in personal contact with other engineers and scientists who were developing new technology. The Wright brothers, who developed their own technological advancements, were in constant contact with Octave Chanute, considered to be the leader in aeronautical technology at the turn of the century, and to some extent with Samuel Langley, who had carried out meticulous aerodynamic experiments in the 1890s. Beginning with the early 1900s, all the remaining grand designers in this book benefitted from contacts with, and membership in, the Royal Aeronautical Society in England and/or the Institute of the Aeronautical Sciences in the United States – two professional societies whose primary purpose was the

dissemination of technical information. The grand designers knew and appreciated what was happening in their technological world. But they also knew how to pick and choose.

(4) The airplanes designed by the grand designers reflected the creativity and knowledge of the *individual person*, not that of a whole team of people. The Wright brothers did their own thing. Although Frank Barnwell and the others considered here worked for airplane companies and had the benefit of teams of helpers, such as an aerodynamics group, structures group, engine group, stability and control group, etc. – sometimes with only one or two people in a group – each grand designer made the important decisions and created his design out of whole cloth. The airplanes designed by Frank Barnwell, R. J. Mitchell, and the others were a final product of that one designer's creativity, in the same sense that Monet created his waterlily paintings, and van Gogh his sunflowers. Indeed, I feel that airplane designers are influenced by aesthetic considerations, albeit in the subconscious, as well as technical considerations.[1,2] C. P. Snow comments on the aesthetic nature of airplane design in his seminal book *The Two Cultures*:

> The ... line of argument draws, or attempts to draw, a clear line between pure science and technology (which is tending to become a pejorative word). This is a line that once I tried to draw myself: but though I can still see the reasons, I shouldn't now. The more I have seen of technologists at work, the more untenable the distinction has come to look. If you actually see someone design an aircraft, you can find him going through the same experience – aesthetic, intellectual, moral – as though he was setting up an experiment in particle physics.

The aeronautical historian, Richard Smith, in an essay dealing with airplane technology, wrote:[3]

> Instead of a palette of colors, the aeronautical engineer has his own artist's palette of options. How he mixes these engineering options on his technological palette and applies them to his canvas (design) determines the performance of his airplane.

[1] Anderson, John D., Jr., "Design for Performance: The Role of Aesthetics in the Development of Aerospace Vehicles," in Anthony M. Springer (ed), *Aerospace Design: Aircraft. Spacecraft and the Art of Modern Flight*, Merrell Publishing Company, London, Chapter 3, 2003.

[2] Snow, C. P., *The Two Cultures: and a Second Look*, Cambridge University Press, 1963.

[3] Smith, Richard, "Better: The Quest for Excellence," in John T. Greenwood (ed.), *Milestones of Aviation*, New York, High Lauter Levin Associates, 1989, p. 243.

> When the synthesis is best it yields synergism, a result that is dramatically greater than the sum of the parts. This is hailed as "innovation." Failing this, there will result a mediocre airplane that may be good enough, or perhaps an airplane of lovely external appearance, but otherwise an iron peacock that everyone wants to forget.

Along these lines, we can single out a major characteristic of the "grand designers." They were in their own right, consummate artists, and their airplanes were literally works of art.

There are two areas in which the grand designers are dissimilar. The first is formal education. There is no correlation whatsoever between the amounts of formal education received by a given grand designer and his design accomplishments. The Wright brothers never officially graduated from high school, let alone went to college. Yet, they invented history's first practical airplane. Frank Barnwell, by going to classes at night, ultimately earned a bachelor's degree in naval architecture. Art Raymond, in contrast, earned both a bachelor's and master's degree in aeronautical engineering from the nascent program at MIT in the early 1920s. R. J. Mitchell had a British technical school education, but never went to college. In total contrast, however, was Kelly Johnson, with a bachelor's and master's degrees from the University of Michigan, one of the best aeronautical engineering programs in the country, then as now. Johnson then went on, while he was working at Lockheed, to become an "all but dissertation" Ph.D. student at the California Institute of Technology in the late 1930s, completing all the required course work for a doctorate, but stopping short of completing his Ph.D. because he did not want to devote the time necessary to satisfy the foreign language requirement for the degree. Of all the grand designers, Kelly Johnson did benefit from his college education because he worked later in the twentieth century, when aeronautical engineering became much more scientifically advanced and technically complex. Johnson's education prepared him to meet the challenges of the second design revolution in the jet age. Johnson, in turn, loved learning, and his advanced education was a natural fit for him. Throughout the first half of the twentieth century, however, there is no evidence that a formal education was a necessary prerequisite for becoming a "grand designer."

The second area of dissimilarity between the various grand designers was the amount of formal publishing they did in the technical archive literature. The Wright brothers, Art Raymond, R. J. Mitchell, and Edgar Schmued published very little if at all, in the literature. The practice of airplane design, indeed, inhibits such publication. Grand designers developed winning techniques and approaches towards their designs, and in the competitive world of new airplane design, there is an understandable reluctance to share this intellectual property with the rest of the world. Also, the productivity and success of airplane designers is measured by their airplanes, not by a list of technical journal papers. On the other hand, Frank Barnwell and Kelly Johnson, on extreme ends of the historical development of airplane design in the first half of the twentieth century, published copiously. Barnwell used the British periodicals *The Airplane* and *Flight*, and later the *Aeronautical Journal* of the Royal Aeronautical Society to publish numerous papers on details of airplane design and on various aspects of aeronautical technology, beginning as early as 1914. Kelly Johnson, who later became a fellow of the Institute of the Aeronautical Sciences (IAS), favored publication in the Institute's *Journal of the Aeronautical Sciences,* America's premier aeronautical technical journal. Indeed, Johnson lost no time after he had graduated from Michigan and started work for Lockheed, publishing his first IAS paper just two years after he joined the company. He published five more IAS papers before the beginning of World War II and then continued with a detailed, almost mind-boggling fifty page paper in the Journal in 1947 on details of his design of the P-80, America's first successful jet fighter. Even the super secret SR-71 Blackbird was the subject of one of his later papers, published by the American Institute of Aeronautics and Astronautics (successor to the IAS). Kelly Johnson felt a compelling desire to help his colleagues in the profession of aeronautical engineering by publishing various "lessons learned," particularly on the design of pioneering new airplanes. Besides, he clearly loved the process of writing technical papers or else he would not have done it. Publishing was not a requirement for his job at Lockheed.

The practice of airplane design changed considerably after Kelly Johnson. New airplanes, such as the Boeing 707 swept-wing transport and all its followers, became complex machines. Their design was out of the hands of just one "grand designer." Rather, they were the products of design teams. Hence, it is not practical to continue with further case histories in this book. Furthermore, the advent of the modern high-speed digital computer makes it possible to consider hundreds, even thousands of different design permutations in the conceptual design process. However, the basic methodology for conceptual airplane design as first outlined by Frank Barnwell a hundred years ago, remains buried in the logic used by the computer programs. It has to. The methodology is based on fundamental laws of nature and the computers are programed to deal with these same laws.

Perhaps the current situation in airplane design is best summarized by Ben Rich, who took over from Kelly Johnson as director of Lockheed's Skunk Works. Rich oversaw the design of another famous Skunk Works airplane, the F-117 stealth fighter. In 1994, Rich wrote:[4]

> In my forty years at Lockheed I worked on twenty-seven different airplanes. Today's young engineer will be lucky to build even *one*.

Such an atmosphere is not fertile ground for the growing of new "grand designers" in the future. Nevertheless, if history has shown us anything, in our modern world there will continue to be new and improved airplanes; it is hard to imagine anything different. In turn, there will continue to be new generations of aeronautical engineers who will play a role in designing these new airplanes. Among them will be new "grand designers."

Finally, the six designers discussed in this book were somewhat arbitrarily chosen as classic case histories strictly for the purpose of investigating and establishing the main findings of this study. Over the past century there have been many hundreds of new and successful airplanes designed by dedicated, innovative aeronautical engineers who in their own right are grand designers. This book is dedicated to those people.

[4] Richard and Janos, p. 316.

Appendix A

Calculation of Change in Maximum Velocity of the Spitfire

Elliptic Wing versus a Straight Tapered Wing

When an airplane is in steady, level flight at some velocity, V_∞, the power available from the power plant, P_A, is equal to the power required to overcome the drag of the airframe at that velocity, P_R (see Anderson, *Introduction to Flight*, 8th ed., McGraw-Hill, 2016, pp. 461–470). The power required is

$$P_R = D\,V_\infty \tag{1}$$

where D is the aerodynamic drag. In turn the drag is given by

$$D = \tfrac{1}{2}\,\rho_\infty V_\infty^2\, S\, C_D \tag{2}$$

where S is the wing planform area, ρ_∞ is the density of the surrounding air, and C_D is the drag coefficient. The drag coefficient is the sum of the parasite drag coefficient, $C_{D,e}$ and the induced drag coefficient, $C_{D,i}$, which is caused by the vortex drag associated with vortices emanating from the wing tips.

$$C_D = C_{D,e} + C_{D,i} \tag{3}$$

The induced drag coefficient is a function of the lift coefficient, C_L;

$$C_{D,i} = \frac{C_L^2}{\pi e AR} \tag{4}$$

where AR is the wing aspect ratio and e is the span efficiency factor. The lift coefficient is defined as

$$C_L = \frac{L}{\frac{1}{2}\rho_\infty V_\infty^2 S} = \frac{W}{\frac{1}{2}\rho_\infty V_\infty^2 S} \tag{5}$$

where in steady, level flight, the lift is equal to the weight. Substituting Equation (5) into Equation (4), and the result into Equation (3), we have

$$C_D = C_{D,e} + \frac{C_L^2}{\pi e AR} = C_{D,e} + \frac{W^2}{\frac{1}{4}\rho_\infty^2 V_\infty^4 S \pi e AR} \tag{6}$$

Substituting Equation (6) into Equation (2),

$$D = \tfrac{1}{2}\rho_\infty V_\infty^2 S\, C_{D,e} + \frac{W^2}{(\frac{1}{2}\rho_\infty V_\infty^2 S \pi e AR)} \tag{7}$$

Substituting Eq. (7) into Eq. (1), and recalling that $P_R = P_A$, we have

$$P_R = P_A = \tfrac{1}{2}\rho_\infty V_\infty^3 S\, C_{D,e} + \frac{W^2}{\frac{1}{2}\rho_\infty V_\infty S \pi e AR} \tag{8}$$

For an airplane with given P_A, wing area S, parasite drag coefficient, $C_{D,e}$, aspect ratio, AR, flying at a given altitude, hence given air density ρ_∞, Equation (8) is simply a relation between flight velocity V_∞ and the span efficiency factor e. Using differential calculus, if we differentiate Equation (8) for V_∞ with respect to e holding all other variables constant, i.e., from the partial derivative $\frac{\partial V_\infty}{\partial e}$, we obtain

$$\frac{\partial V_\infty}{\partial e} = \frac{V_\infty}{2e} \tag{9}$$

Taking the second partial of Equation (9) with respect to e,

$$\frac{\partial^2 V_\infty}{\partial e^2} = -\frac{V_\infty}{2e^2} \tag{10}$$

The change in flight velocity, ΔV_∞ for a given change in e, Δe, can be obtained from a Taylor's series expansion:

$$\Delta V_\infty = \frac{(\partial V_\infty)}{\partial e}\Delta e + \frac{\partial^2 V_\infty}{\partial e^2}\frac{(\Delta e)^2}{2} + \text{High order terms} \tag{11}$$

For the Mark I version of the Spitfire (the first production model), the maximum velocity, V_{max}, was 362 mph at 18,500 feet. Also, for the elliptic wing of the Spitfire, e = 1. Thus, from Equations (9) and (10),

$$\frac{\partial V_{max}}{\partial e} = \frac{362}{2(1)} = 181 \tag{12}$$

and

$$\frac{\partial^2 V_{max}}{\partial e^2} = -\frac{362}{2\,(1)^2} = -181 \tag{13}$$

Examining the straight tapered wing of the initial design for the Spitfire (Figure 5.12a), the taper ratio (tip chord/root chord) is 0.57. The aspect ratio of the Spitfire Mark I is 5.6. Using the straight tapered wing with the same aspect ratio of the Spitfire's elliptical wing, from Anderson, *Fundamentals of Aerodynamics*, 5th Edition, pages 437–440, we find the value of the span efficiency factor for the non-elliptical wing to be $e = 0.985$. Hence, for the design in Figure 5.12a compared to the final Spitfire design in Figure 5.12d, the value of e is reduced from 1 to 0.985, and therefore in Equation (11), we have $\Delta e = 0.985 - 1 = -0.015$. Inserting this value for Δe and the values for the derivatives from Equations. (12) and (13) into Equation (11), we have

$$\begin{aligned} \Delta V_{max} &= \frac{\partial V_{max}}{\partial e}\Delta e + \frac{\partial^2 V_{max}}{\partial e^2}\frac{(\Delta e)^2}{2} + \dots \\ &= (181)(-0.015) + (-181)((-0.015)^2)/2 + \dots \end{aligned}$$

or,

$$\Delta V_{max} = \boxed{-2.74\ mph}$$

Conclusion: By using the straight tapered wing shown in Figure 5.12a rather than the elliptic wing shown in Figure 5.12d, the maximum velocity of the Spitfire Mark I would have been reduced by only 2.74 mph.

Appendix B

Thrust Available

Comparison of a Piston Engine/Propeller Combination with a Jet Engine

Using the methodology from Chapter 6 of Anderson, *Introduction to Flight,* 8th Edition, McGraw-Hill, New York, 2016, we calculate the thrust available from an Allison V-1710-11 piston engine/propeller combination used by Kelly Johnson for the Lockheed P-38L Lightening and the Allison J33-A-11 jet engine used by Johnson on the Lockheed P-80A, both at an altitude of 20,000 feet.

PISTON ENGINE

The Allison V-1710-11 piston engine has a power output at sea level of 1,470 horsepower. On the P-38, the engine was turbosupercharged, providing the same power output at 20,000 feet. Letting η represent the propeller efficiency and P the shaft power output from the engine, the power available, P_{avail}, from the engine/propeller combination is (Anderson, p. 465):

$$P_{avail} = \eta P \tag{1}$$

Dealing with consistent units of foot-pounds per second for power, we note that

$$550 \text{ ft-lb/ sec} = 1 \text{ horsepower}$$

Hence, from Equation (1), assuming a propeller efficiency of $\eta = 0.82$, we have

$$P_{avail} = \eta P = (0.82)(1470)(550) = 6.6297 \times 10^5 \text{ft-lb/ sec}$$

The power available is also given by the product of the flight velocity, V_∞ and the thrust available, T_{avail}. (Anderson, p. 468)

$$P_{avail} = V_\infty T_{avail} \qquad (2)$$

or,

$$T_{avail} = \frac{P_{avail}}{V_\infty} \qquad (3)$$

Equation (3) allows the calculation of thrust available from a reciprocating engine/propeller combination. Note that the thrust available is a function of flight velocity, which for this calculation is chosen as 420 mph. Equation (3) requires consistent units, which for velocity are ft/sec. Using the conversion factor that 88 ft/sec = 60 mph, we have

$$V_\infty = 420\left(\frac{88}{60}\right) = 616 \text{ ft/ sec.}$$

From Equation (3), we have for a flight velocity of 420 mph,

$$T_{avail} = \frac{P_{avail}}{V_\infty} = \frac{6.6297 \, x \, 10^5 \, ft\text{–}lb/\text{ sec}}{616 \, ft/\text{ sec}} = \boxed{1076 \, lb}$$

JET ENGINE

For the Allison J-33-A-11 jet engine used on the P-80A, the sea level thrust is T_o = 4000 lbs. The thrust of a turbojet engine varies directly with the density of the ambient air; hence, it decreases with increasing altitude. (Anderson, p. 471.) The air density at a standard altitude of 20,000 ft is 1.2673×10^{-3} slug/ft^3 (Anderson, p. 875). The air density at sea level is 2.377×10^{-3} slug/ft^3. Thus, at 20,000 ft., the thrust from the Allison J33-A-11 is

$$T = 4000\left(\frac{1.2673x \, 10^{-3}}{2.377 \, x \, 10^{-3}}\right) = 4000(0.533) = \boxed{2132 \, lb}$$

COMPARISON

Clearly at an altitude of 20,000 feet, the jet engine provides almost *twice as much thrust* as does the reciprocating engine/propeller combination.

(Note: This is a comparison between the two types of engines, one-to-one, for the purpose of contrasting the output of a single reciprocating engine/propeller combination with that from a single jet engine. The Allison reciprocating engine was chosen as an example because it was the choice made by Kelly Johnson for the P-38. The fact that the P-38 used two of these engines does not impact the comparison made here.)

Index

Printed in the United States
by Baker & Taylor Publisher Services